D1448838

SCIENCE

A

New York Times

SURVEY

SCIENCE

A
New York Times
SURVEY

Edited by Arleen Keylin

Introduction by Walter Sullivan

VOLUME
1

Arno Press

New York • 1980

61491

Copyright © 1980 by Arno Press Inc.
Copyright © 1976, 1977, 1978, 1979 by The New York Times Company.
Reprinted by permission.

Library of Congress Cataloging in Publication Data

Main entry under title:

Science, a New York times survey.

    Includes index.
    1. Science—Addresses, essays, lectures.
I. Keylin, Arleen. II. New York times.
Q171.S3374   500   80-26178
ISBN 0-405-13970-5 (set)

Book Design: Lynn Yost

Assistant Editor: Douglas John Bowen

Manufactured in the United States of America

Contents

Volume 1

BIOLOGY

Volume 2

TECHNOLOGY

HEALTH AND MEDICINE

Volume 3

ARCHEOLOGY

Volume 4

ASTRONOMY AND SPACE

THE NOBEL PRIZES IN SCIENCE

CHRONOLOGY OF SCIENCE EVENTS

Contributors

Lawrence K. Altman, M.D., is a reporter for *The New York Times.*
Scientists Brave Bears and Winter in Hunt for a Sleep Hormone
Penicillin
1979 Nobel Prizes in Science—Physiology and Medicine

Alan Anderson Jr. is editor of *The Illnois Times,* a weekly newspaper published in Springfield.
Barry Commoner: Scientist-at-Large

Leslie Bennetts is a reporter for *The New York Times.*
Biofeedback: From Counterculture to Fad to Accepted Medical Tool

Colin Blakemore is fellow and director of studies in medicine, Downing College, Cambridge.
The Unsolved Marvel of Memory

Edward M. Brecher is a longtime writer on science and medicine.
Opting for Suicide

Jane E. Brody is a reporter for *The New York Times,* and
co-author of *You Can Fight Cancer and Win.*
Researchers Seek New Male Contraceptive
The Countless Mysteries of Peatland
Inbred Mice
Plant-Animal Interaction

Blanche R. Brown is a professor of fine arts at New York University and author of *Free Cities,*
an art guide to Athens, Rome, Florence, Paris and London.
Out of the Ashes: Glowing Treasures of Pompeii

Michael H. Brown is a reporter for *The Niagara Falls Gazette*.
Love Canal, U.S.A.

Malcolm W. Browne is a reporter for *The New York Times*.
Scientists Expect New Clues to Origin of the Universe
Nobel Laureate Seeking to Unravel the Shape of an Enzyme
The Solar Eclipse
Man-Powered Flight
1978 Nobel Prizes in Science—Physics
1979 Nobel Prizes in Science—Chemistry and Physics

Tom Buckley is a reporter for *The New York Times*.
Is Anybody Out There?

William E. Burrows is an associate professor of journalism at New York University.
Cancer Safety

Gilbert Cant is the medical editor of *The New York Times*.
Worrying About Ulcers

Douglass Cater was a long-time Washington writer and is now president of *Observer International* in London.
How to Have Open Heart Surgery (And Almost Love It)

Liebe F. Cavaliere is a member of Sloan-Kettering Institute for Cancer Research and professor of biochemistry at Cornell University Graduate School of Medical Sciences.
New Strains of Life or Death

Laurence Cherry writes frequently about medicine and science.
Solving the Mysteries of Pain
A New Vision of Dreams
Anesthesiologists
Medical Technology: The New Revolution
New Hope for Burn Victims

Barry Commoner is director of the Center for the Biology of Natural Systems at Washington University, St. Louis. He is chairman of the board of directors of the Scientists' Institute for Public Information and most recently, the author of *The Poverty of Power*.
The Promise and Perils of Petrochemicals

Glenn Collins is an editor of *The New York Times Magazine*.
Life With Father

Patricia Curtis is a freelance journalist who writes frequently about animals, and has completed a book about animal rights.
Animal Experiments
Pet Therapy

Lee Edson writes frequently about science for *The New York Times*.
The Advent of the Laser Age

Timothy Ferris is a professor of English at Brooklyn College and author of *The Red Limit: The Search for the Edge of the Universe* and co-author of *Murmurs of the Earth: The Story of the Voyager Interstellar Record*.
Seeking an End to Cosmic Loneliness
Mysteries of Deep Space
Crucibles of the Cosmos

Franklin Folsom, a freelance writer, is author of *America's Ancient Treasures* and *Science and the Secret of Man's Past.*
'Space-Age' Archeology

C.P. Gilmore is executive editor of *Popular Science* magazine and science editor of WNEW-TV.
Taking Exercise to Heart

Wade Green writes frequently on social and environmental issues.
The New Alchemists

Alexis Greene is a freelance writer working on a book about television and the arts.
The Coming Impact of Technology on the Arts

Michael Harwood is a Connecticut writer.
The Rising Tide of Oil Spills

Robin Marantz Henig is a features and news editor of *BioScience Magazine* in Washington, D.C.
The Myth of Senility
Mother's Milk

Gladwin Hill is a correspondent to *The New York Times* on environmental issues.
Restrictions Are Impeding the Use of Solar Devices

Janet L. Hopson is a science writer.
A Plea for a Mundane Mollusk

Evan Jenkins is an assistant news editor for *The New York Times.*
Computers Programmed for a Revolution

Elisabeth Keiffer is a freelance writer and editor of a bimonthly newsletter covering marine subjects in New England and New York.
Salmon Ranches and Lobster Farms

Grace Lichtenstein, a *New York Times* correspondent, is the author of *Desperado,* a book about her experiences in the West.
Battle Over the Mighty Colorado

Richard D. Lyons is a correspondent for *The New York Times.*
Scientists Discover New Form of Life that Predates Higher Organisms
Skylab

D.H. Melhem, author of *Rest in Love,* is a New York poet.
A Family Miracle

Anthony J. Parisi is a business reporter for *The New York Times*.
The Energy-Efficient Society

Donald R. Pellman is a freelance writer.
Learning to Live with Dying

Maya Pines frequently writes about medicine and psychiatry for *The New York Times Magazine*.
St-st-stuttering
Heredity Insurance

Boyce Rensberger was a reporter for *The New York Times*.
The World's Deserts Are Growing
Pennsylvania Site Yields Clues
The Wolf Gets a Better Image
The World's Oldest Works of Art
Ice Age Tools Put to the Test
This Is Not the End of the Game
The Ebla Ruins
Rival Anthropologists

Cokie Roberts is a freelance journalist. **Steven V. Roberts** is chief of
The New York Times Athens bureau.
Atlantis Recaptured

Hope Ryden is the author of *God's Dog*, a book about coyotes.
The Clamor to Poison the Coyote

Carl Sagan is David Duncan Professor of Astronomy and Space Sciences and director of the
Laboratory for Planetary Studies at Cornell University. He is the author, most recently, of *Broca's
Brain* and *Cosmos*.
The Next Great Leap into Space

Maggie Scarf has just completed a book published by Doubleday, on stress
points in the lives of women.
From Joy to Depression
Shocking the Depressed

Harold M. Schmeck Jr. is a reporter for *The New York Times*.
The Key to Legionnaire's Disease
Searching for Insulin's Matching Lock in Diabetes
Smallpox Virus Is Stored in Lab for Insurance
Scientists Seek Clues to Nervous Disorders
1978 Nobel Prize in Science—Medicine
1978 Novel Prize in Science—Chemistry
The Manic-Depressive Cycle
Aging

Dodi Schultz is co-author of *The Mothers' and Fathers' Medical Encyclopedia*.
Why Childhood Disease Are Coming Back

Richard Severo is a reporter for *The New York Times*.
Too Hot to Handle

Philip Shabecoff is a *New York Times* correspondent.
New Battles Over Endangered Species

Kathy Slobogin works for the editorial staff of *The New York Times*.
Stress

Richard Spark, M.D., is associate clinical professor of medicine at Harvard Medical School and Beth Israel Hospital in Boston.
Breaking the Drug Barrier

William Stockton is director of science news and editor of *Science Times* at *The New York Times*.
Smallpox Is Not Dead
Pilots Talk About Air Crashes
A New Clue in the Cancer Mystery
Dual Addiction

Walter Sullivan is science editor of *The New York Times*,
and author of *Black Holes: The Edge of Space, the End of Time*.
Uranus Has Five Rings
Australia's Biosphere Reserves
Genetic Decoders
New Clues to Animals of the Ice Age
Balloon Data Indicate Vast Amounts of Antimatter Are in Milky Way
Toward Harnessing Stars' Energy
An Astronomer Looks Back to See Ahead
A World Astronomy Center
Two Scientists at Field Station in Panama Study Malaria-Bearing Lizards Unharmed by Disease
Scientists Struggle to Preserve Hawaii's Strange Species
Radar Maps a Rugged Land Beneath Antarctica
Scientists Praise British Birth as Triumph
Fusion
The Elephant Seal

Ludmilla Thorne is a freelance writer who specializes in Russian subjects.
Inside Russia's Psychiatric Prisons

Bayard Webster is a reporter for *The New York Times*.
The Biggest Dinosaur

William Wertenbaker, author of *The Floor of the Sea*, is a staff writer at *The New Yorker* and is now working on a book about early man.
Mining the Wealth of the Ocean

Marietta Whittlesey is a freelance writer and author of *Killer Salt*.
Antibiotics

John Noble Wilford was the director of science news at *The New York Times*.
Shuttling Through Space
New Images of Earth from Space
The Spinoff from Space
'Ship' 500 Miles Out in Space to Explore Oceans of World
Venus Probes Yield Mass of Data

Roger Neville Williams is a freelance writer living in Telluride, Colorado.
Tiny Town vs. Mining Giant

Anthony Wolff is a New York City journalist who often writes on country matters.
Building a Better Bug Trap.

Introduction

by Walter Sullivan

It can be said that in 1977 mankind looked farther into the distance, into the past, into the inner workings of the atom and of the life process than ever before. It would probably be possible to say the same of 1978 and 1979, at least in many respects, for the current advances in human knowledge are like a great river, moving inexorably on.

A major reason for this flow is the development of new ways to observe, be it with spacecraft, novel telescopic systems, atom smashing devices (some several miles in circumference) and ingenious ways to decipher the structure of such critical components of life as the very long molecules of DNA (deoxyribonucleic acid). They include, as well, the bold use of manned vehicles to explore regions otherwise denied human eyes, inaccessible mountains of Antarctica, rifts and metal-erupting geysers in the sea floor, as well as unmanned vehicles to survey other planets.

The year 1977 was marked by a growing awareness that mankind, in terms of population growth and demands for raw materials and energy, has begun pressing the limits of the earth's resources. The population in much of the world has become so dense and air travel has so weakened the traditional barriers against disease spread that the appearance of any new organism against which the population has no defenses is a cause for grave concern. Early in the year identification of the cause of Legionnaire's disease showed it to be a previously unrecognized organism that killed many of those whom it infected. Fortunately it was not highly contagious, although during the year it reappeared in a few isolated instances. Had it occurred in epidemic form the results could have been disastrous.

The range of human vision has been greatly widened by a new generation of telescopes and the development of electronic viewing systems, assisted by computer, that tell things about extremely distant observable objects that could not be deduced from photographs.

Thus, exploration of our world, of the life upon it, past and present, of the planet's neighbors in the solar system, and of the universe that is its home proceeds. While 1977 saw advances on numerous fronts, it also set the stage for new discoveries, new surprises, new insights bound to follow in the years thereafter.

INTRODUCTION

The year 1978 will be remembered for a number of things—certainly for the birth of the first baby conceived in the laboratory. But a generation may pass before it becomes clear which other developments were historic. For example. what seems a major advance toward a vaccine against malaria was reported. If it matures into an effective weapon against that disease it will save at least a million lives per year in areas, such as India, Africa and South America, where the disease has again become rampant.

Other important steps were taken toward harnessing the nuclear reaction that makes the sun shine. Have they really brought closer the day when unlimited energy can economically be derived from sea water? This may not be clear until the end of the century.

Our spacecraft reached far into space, probing through the clouds of Venus, and demonstrated new ways of looking at the earth's surface, notably to record sea surface conditions worldwide. While the controversy over genetic manipulation simmered, its potential was demonstrated in the transfer to bacteria of the genetic signals for production of such substances, critical to human well being, as insulin.

The year 1979 was memorable for developments applicable to such diverse questions as the essential character of nature's laws, the well-being of mankind and the ultimate fate of the universe.

It was a year for looking back—at the revolutionary contributions of Albert Einstein, born a century earlier—and ahead toward the possible fulfillment of his dream, incorporating all natural phenomena into an integrated fabric of understanding.

1979 was a year when, because of a major nuclear accident and increasing concern at the affect on climate of heavy fuel burning, the prospects for an easy solution to the energy crisis became even more remote.

With more pedestrian forms of exploration obsolete, attention turned upward, as some of the most dramatic space discoveries to date revealed the amazing diversity of Jupiter's moons, and downward to the deep sea floor where remarkable hot-water geysers were observed in action.

In medicine the day drew nearer when genetic manipulation could be used to deal with genetic disorders and fears for the hazards of such research diminished. Further evidence was found that the brain produces its own pain-killers.

□

Was 1979 more productive, scientifically, than other years? One cannot judge from so short a perspective. But new tools of research, such as those in biomedical laboratories for transferring genes or the most powerful particle accelerators and orbiting spacecraft, are opening new frontiers so rapidly that the momentum is likely to sustain such advances into what may become a period of considerable budgetary austerity.

ENERGY AND ENVIRONMENT

With regard to great natural catastrophes that threaten our lives, such as earthquakes and radical climate changes, there were new developments but no definitive breakthroughs. The frigid winter of 1976-1977 led to cries that an ice age is upon us, but there have been colder winters in the past, when no ice age followed. A number of ice age theories are being discussed, including one that blames intensified volcanic activity, loading the stratosphere with ash particles that cut off sunlight. The most widely accepted theory puts the blame on cyclic changes in the earth's orbit around the sun and the tilt of its spin axis, periodically reducing the amount of solar energy reaching high latitudes. According to that timetable the next ice age is many centuries off and the chief concern to emerge in 1977 was that massive burning of coal during the next century will so raise the level of carbon dioxide in the air as to warm the climate.

President Carter's energy plan called for heavy dependence on coal to avoid excessive development of nuclear energy, but a study for the National Academy of Sciences pointed out that coal burning might release more carbon dioxide than could be absorbed by accelerated plant growth and absorption in the oceans. If that gas remains in the atmosphere it will act like the glass of a greenhouse, which permits entry of sunlight to warm the interior, but inhibits the escape of heat in the form of infrared radiation. The effect on climate would be complex and could be disastrous for food production that already is barely able to stave off famine in some regions.

An initial step was taken in 1977 to head off an adverse effect on climate and public health when, as of November 1, the Food and Drug Administration required that spray cans using chlorofluorocarbons as propellants (typically to dispense hair sprays, deodorants and other toiletries) bear a warning label stating they "may harm the public health and environment." It is hoped use of such propellants can be phased out entirely, lest they deplete stratospheric ozone that shields the earth from the ultraviolet component of sunlight causing skin cancer. It is suspected that the shielding process also affects climate. Such propellants, best known under their DuPont trade name of Freon, are ideally suited as propellants because, even under the great pressure in aerosol cans, they do not affect chemically whatever is to be ejected, be it hair spray, deodorant or some other cosmetic preparation. However, there is evidence that, when the fluorocarbons reach the stratosphere, they break down, under ultra-violet exposure, releasing chlorine that then depletes the ozone. Fluorocarbons had been used in many—but not all—spray cans and are the standard refrigerant in air conditioning systems. Limiting their escape from such systems will be far harder to control.

The most spectacular event, affecting climate, would be a large-scale surge of Antarctic ice. Such surges are known to occur where a glacier has become unstable and begins flowing rapidly. Contributing factors are the configuration of the rock basement beneath the ice and the extent of ice accumulation. The ice sheet covering West Antarctica, a mile or more in thickness, rests on land that is largely below sea level and, from drilling and radar probing, it appears that the ice rests on a layer of slush that could permit rapid dispersal of the ice into icebergs. Between the last two ice ages the world oceans were about 30 feet higher than today and this has led some glaciologists to suspect that the West Antarctic ice sometimes slips into the sea. In the southern summer of 1976-1977 an international research project was carried out to explore further this possibility, particularly in terms of evidence lying within and beneath the Ross Ice Shelf. The latter is an apron of continental ice floating on the waters of an embayment the size of France. It lies in the path that would be followed by such a surge and may play a controlling role. The project is continuing.

For Californians the year was marked by continuing concern about the possibility of a major earthquake, particularly to the northeast of Los Angeles. In that area a section along the San Andreas Fault had risen close to a foot as though in an accumulation of strain prior to release in a great quake. In 1977 some of this swelling—"the Palmdale bulge"—subsided. While a variety of premonitory signs that could warn of an impending quake have been identified, their cause and meaning are not sufficiently understood for reliable predictions. The Chinese, faced with the most devastating quakes in the world, are using all suspected warning signs, from sophisticated geophysical measurements to the unusual behavior of animals, and have made one dramatically successful prediction. In February 1975, they ordered the evacuation of Haicheng, a populous community, and a quake ensued sufficiently severe to have taken thousands of lives. Unfortunately, a year later, a far worse quake occurred in Tangshan which is believed to have taken 650,000 lives. Some warning signs were observed, according to reports published in 1977, but either they were ambiguous or else political power struggles in Peking led to no action being taken.

Another successful prediction, however, was made on November 2, 1978, a few hours before a severe earthquake that shook a remote region of the Alai Mountains in the Soviet Union. They lie near the border between Tadzhikstan and Kirghizia. It is a region where some 12,000 people were entombed by landslides following a quake in 1949 and, as a result, intensive efforts at prediction have been under way there. While a variety of warning signs were recorded before the November quake it was the slowing or cutting off of water flow from eleven artesian wells in the area on November 1 that led to the warning.

In the United States some earth scientists expressed concern at the presence of a seemingly active geologic fault, one branch of which passes within 3,000 feet of the nuclear power plants at Indian Point on the Hudson River. They predicted that there was a 5 to 11 per cent chance that an earthquake might occur along that fault exceeding in magnitude the design limits of the plant.

On a national level President Carter on June 22 transmitted to Congress a plan for a National Earthquake Hazards Reduction Program, prepared at the request of Congress.

Among the most surprising discoveries to become known in 1978 was a remarkable combination of regularity and irregularity in the sunspot cycle. It has long been known that the number of sunspots reaches its maximum roughly every eleven years, but the timing is irregular, the intervals ranging from seven to seventeen years. Since eruptions ("solar flares") occur near sunspots, such solar activity follows a similar cycle. The eruptions eject great jets of magnetized gas into space which reach the earth some days later. Flashes of X-rays associated with the flares also affect ionization (electrification) of upper atmospheric layers on earth, cutting off long-range radio communications.

The sunspots are strongly magnetized, those in one hemisphere being of opposite magnetic polarity from those on the opposite side of the solar equator. With each new cycle this magnetic polarity flips over, so the full magnetic cycle (when polarity returns to its previous configuration) takes some twenty-two years.

The first clue to a more fundamental irregularity in the sun's behavior—one that seems to have an important effect on climate—was the documentation by Dr. John A. Eddy of the High Altitude Observatory in Boulder, Colorado, of a time from 1645 to 1715 when virtually no sunspots at all were visible. This was the period of "The Little Ice Age" when winters were long, summers short, crops failed and there was great suffering in much of the world. Eddy and his colleagues then found, from records kept by early astronomers, that as that hiatus in sunspot activity, known as the Maunder Minimum, approached, the sun's rotation rate speeded up, primarily near its equatorial region. The sun, being a gaseous body, does not rotate as do solid objects. At present a sunspot near its equator rotates every 25 days. Near the poles the rate is 34 days.

From records kept by Christopher Scheiner from 1625 to 1626 and by Johannes Hevelius from 1642 to 1644 it appears that there was a speed-up of from 3 to 5 per cent in the rotation rate near the solar equator as "The Little Ice Age" drew near, the effect being less marked at higher latitudes. Richard B. Herr of the University of Delaware then studied unpublished drawings on sunspot locations made in England by Thomas Harriot from 1611 (soon after invention of the telescope) until 1613. In thus extending the record even farther back Herr was able to show the speed-up was even more marked than had been believed. The implication is that recording a contemporary increase in spin rate could be taken as warning of reduced solar activity and climate cooling.

Juxtaposed against these observations was a finding by Dr. Robert H. Dicke of Princeton University that the solar cycle, despite its manifested irregularity, is controlled by a very precise "chronometer" somewhere deep within the sun. If sunspot maxima are analyzed statistically, it turns out that while they occur irregularly on a scale of decades, they conform to a very precise timer when averaged over the centuries.

This internal "heart" pounds away at a pulse rate of once every 22.27 years, give or take a month, that being the period for the full magnetic cycle. Furthermore it seems to match cycles in climate reflected in the ratio of deuterium (the heavy form of hydrogen) to normal hydrogen in tree rings. This ratio, in rain and snow, appears to be controlled by atmospheric and sea surface temperatures. The precipitated moisture is then taken up by trees, such as the long-lived bristecone pines of the Sierras. Scientists from the California Institute of Technology have traced this record of climate back 1,000 years by analysis of the wood in successive tree rings and Dr. Dicke has found that the cycle in climate seems to follow the pulse rate embedded within the sun.

In a parallel development Dr. Bruce C. Parker and his colleagues of Virginia Polytechnic Institute in Blacksburg reported that in the layers of ice at the South Pole Station they had found what seems a record of solar activity reaching back to A.D. 1150 with a potential, by boring deeper, of

extending it several hundred thousand years. They believe the abundances of two nitrogen compounds (ammonia and nitrate) reflect the level of solar activity at the time that snow fell at the pole and, over the years, was buried by more snow and compressed into ice, capturing bubbles of the contemporaneous atmosphere. One explanation for the seemingly increased abundance of the nitrogen compounds in layers formed during a sunspot maximum could be that electrical effects caused by solar activity, such as auroral displays, helped form ("fix") additional nitrogen compounds. Dr. Parker and his coworkers believe that, in their record, they see the effects of the Maunder Minimum that caused a "Little Ice Age" in warmer latitudes.

The year 1978 was marked by increasing concern over possible climate changes, in part because world food production has become so marginally capable of feeding the world's billions and is so vulnerable, being based on what is clearly an unusually warm period in the earth's history. Records kept atop the great Hawaiian volcano, Mauna Loa, since 1955 show an inexorable increase in the carbon dioxide content of the atmosphere, due clearly to increased fuel burning throughout the world. Carbon dioxide acts in the atmosphere much like the glass in a greenhouse, allowing sunlight to pass and heat the earth and its air, but impeding the escape of heat in the form of infrared rays. The zigzagging pattern of carbon dioxide measurement shows an increase during winter, when plants of the Northern Hemisphere are dormant and not taking up carbon dioxide in their formation of carbohydrates. There is then a dip each spring as they return to life; but the overall trend is a steady climb in atmospheric levels of that gas.

Nevertheless two groups of specialists in 1978 said they saw no signs of an impending, catastrophic change in climate. One report, by a group of American, German and Japanese scientists published in the British journal *Nature,* saw no end in sight for the cooling trend that has affected the Northern Hemisphere over the past thirty years. A poll of climatologists in seven countries, conducted by the National Defense University for various government departments, indicated no expectation of major changes before the end of the century, but suggested that in the next century such factors as the increased carbon dioxide—particularly if coal comes into heavy use, rather than energy sources not dependent on combustion—could lead to more serious problems.

On the energy front steady but slow progress was made in 1978 along the rocky road toward fusion—harnessing the reaction that generates the sun's energy. There were advances on both fronts—"magnetic confinement" and "inertial confinement." The goal, in both cases, is to compress and heat nuclei to fuse into helium nuclei. The helium nucleus is a tiny bit lighter than the combined weight of the original two nuclei (because the binding energy is less) and this leftover "material" is converted into a large amount of energy. This is the energy source of a hydrogen bomb, where the necessary conditions for fusion are produced by an atomic bomb trigger.

The attempt to do this in a controlled manner that has been under way the longest is the use of magnetic fields to compress and confine a gaseous mixture of deuterium and tritium. This is being attempted at a number of laboratories, the most ambitious effort being construction of the Tokamak Fusion Test Reactor at Princeton at a projected cost of $225 million. It is to be completed in 1981. Milestones along this road are marked by a combination of three factors: density of the gas (a fully ionized plasma), its temperature and length of time (measured in thousandths of a second) that it can be kept at such density and temperature. These combined factors must produce enough fusion (releasing high energy neutrons) to generate more energy than goes into the system. Princeton announced an important advance in that direction with its intermediate-sized Tokamak during 1978.

In the inertial confinement approach a tiny pellet of deuterium-tritium fuel is slammed from all sides by some form of radiation to crush it to the required conditions of temperature and pressure. In most cases this is being attempted with converging laser pulses, but electron beams are also being tried and the possibility is being explored of using beams of atomic nuclei. In the laser line of attack the largest scale device is Shiva, a system of twenty converging laser beams that began operation during the year at the Lawrence Livermore Laboratory, operated for the government by the University of California. Its converging pulses carry twenty thousand billion watts slamming the pellet, in a billionth of a second, with twenty-five times more power than flows through all the electric utility lines of the United States combined. Shiva, named for many-armed Hindu god of creation and destruction, was built

at a cost of $25 million.

In the field of solar energy a large field of mirrors was completed at the Sandia Laboratories near Albuquerque New Mexico, each of them controlled to reflect sunlight to a boiler atop an experimental tower. An even larger installation is projected for Barstow, California, but the cost of such installations, per unit of generated power, is very high. The same continues to be true for the direct conversion of sunlight into electricity, although some progress in such "solar cells" was reported during the year.

During May a conference was held in Chicago by the American Society of Mechanical Engineers to discuss the derivation of energy from waste burning. It was noted that in 1975 American cities dumped or buried 130 million tons of waste that, had it been burned (according to the federal Department of Energy), would have reduced American oil imports by 7 per cent. The West German city of Munich, with a population of 1.4 million, is deriving 11.8 per cent of its electricity by burning its own garbage and other wastes, the conferees were told. In France, Paris saves 480,000 barrels of oil yearly by burning 1.7 million tons of its refuse. Such efforts in the United States are on a far more modest scale—in Ames, Iowa and Akron, Ohio for example.

The public furor early in 1978 over mysterious booms heard, particularly, along the East Coast, led to a proposal that the deep earth contains almost unlimited reserves of flammable gas (methane). According to Dr. Thomas Gold of Cornell University such gas was captured within the earth during the planet's original formation and escapes from time to time, causing atmospheric explosions as well as the rumbles and flashes of light associated with great earthquakes such as the one that struck the Haicheng region of China on February 4, 1975.

A number of the booms appear to have been caused by supersonic aircraft, both those operated by the military and Concorde airliners en route to and from Europe. Dr. Gold's hypothesis has won some scientific supporters, but remains controversial.

After years of debate on whether a new ice age is imminent, whether supersonic transports, spray can propellants or carbon dioxide from increased fuel burning will alter the world climate, leading climate specialists from throughout the world met in Geneva during February 1979 to compare notes and chart a "World Climate Program." The latter, under the auspices of the World Meteorological Organization, based in Geneva, is designed to collect enough data on factors determining climate and climate change to make possible at least moderately reliable projections.

It was widely agreed that no clearcut trend can currently be projected. Weather records, chiefly from the Northern Hemisphere, have shown a slight warming from the late nineteenth century to about 1950, followed by a slow drop in mean annual temperatures. In at least parts of the Southern Hemisphere, however, opposing trends were recorded, so the global effect seems to have been minor. Masking such trends were regional extremes such as the paralysis of Chicago by snow shortly before the conference, a report from Russia that the month of December was colder than at any time in a century, and records from Britain and Western Europe showing the summer of 1976 to have been the hottest in 250 years. The meeting coincided with a five-week period of intensified worldwide weather studies forming part of the year-long Global Weather Experiment. Balloons were released from 40 ships strung out along the Equatorial Zone, including thirteen ships from the Soviet Union, six from the United States and two from China. Planes parachuted instruments down through the tropical atmosphere. In the Southern Hemisphere 220 buoys provided by six nations were set adrift. Data from the buoys and balloons were collected by a French data-processing system riding the Tiros-N satellite, placed in an orbit passing near both poles by the United States. Additional oceanic data were collected by 80 specially equipped, wide-bodied transports of ten airlines. It was such intensive collection of data from a large portion of the earth's atmospheric and water envelopes, combined with processing by new generations of computers, that, it was hoped, could serve the projected World Climate Program.

The conference took note of recent suggestions that pollutants from various human activities are depleting the stratospheric ozone layer that protects the earth and its inhabitants from harmful wavelengths of ultraviolet sunlight. Among the pollutants under suspicion are oxides of nitrogen from high-flying aircraft and from heavy use of nitrogen fertilizers as well as synthetic substances (known as chlorofluoromethanes or Freons) used as refrigerants and spray can propellants. The conferees also took note of the concern expressed in the previous year at the increasing use of fuel burning as an energy source, adding to the carbon dioxide content of the atmosphere. Carbon dioxide, as a gas, acts much in the manner of the glass in a greenhouse, allowing sunlight to reach the earth unimpeded but preventing the return of heat into space as infrared radiation. The result, it was assumed, would be to warm world climates. Plants, as they grow, remove carbon dioxide from the air and release oxygen. When plants or plant derivatives (such as petroleum products) are burned, the process is reversed (oxygen is consumed and carbon dioxide released).

In their final declaration the conferees stated that during the past century fuel burning and changes of land use, such as felling forests for new farmland, had increased the carbon dioxide content of the air by 15 per cent. Currently it was rising at 0.4 per cent yearly, the declaration said. Forests remove carbon dioxide from the air and place it in long-term storage. When they are felled, the wood is burned or decays, returning that gas to the air. It "appears plausible," said the declaration, that the projected increase of atmospheric carbon dioxide will lead to a gradual warming of the lower atmosphere, especially at high latitudes. In conclusion the declaration stated: "The long-term survival of mankind depends on achieving a harmony between society and nature. The climate is but one characteristic of our natural environment that needs to be wisely utilized. All elements of the environment interact, both locally and remotely. Degradation of the environment in any national or geographical area must be a major concern of society because it may influence climate elsewhere. "The nations of the world," it continued, "must work together to preserve the fertility of the soils, to avoid misuse of the world's water resources, forests and rangelands, to arrest desertification, and to lessen pollution of the atmosphere and the oceans."

The concern over climate change caused by fuel burning raised grave concerns regarding future American energy policy. The reactor accident at Three Mile Island in Pennsylvania, while threatening a serious release of radiation, was controlled and public health specialists concluded that there had been little or no effect on the surrounding population. Nevertheless, the accident and the highly critical report by the investigation commission that looked into it struck a severe blow at energy. The most readily accessible alternative was the burning of coal and its liquid or gaseous derivatives, but widespread concern developed regarding serious climate changes that might affect global food production, starting some time in the next century.

With energy production by atom splitting in trouble, attention turned to atom fusing—the reaction that powers the sun and hydrogen bombs. Large fusion machines built, with variations, on the Soviet Tokamak principle, were under construction in Britain, Japan and the Soviet Union, as well as at Princeton, and it appeared that most, or all, would "break even" in releasing more fusion energy than is injected into the fusion fuel through magnetic compression. On the basis of these hopes an international consortium of those building these machines was formed to develop the world's first fusion reactor—a machine that would produce energy in usable form. It would be purely experimental and fusion was not expected to make a substantial contribution to energy needs until some time in the next century.

Mindful that the present diversity of life on the earth, the product of millions of years of evolution, is threatened in many ways by human activity, wildlife specialists from 50 nations met in Costa Rica in March to draft regulations for the protection of more than 130 species of plants and animals, ranging from orchids to whales. Researchers at the University of California in Santa Cruz reported that one endangered species, the Northern Sea Elephant, was making a dramatic comeback, but that its genetic diversity had probably been irreparably damaged when its numbers were reduced to a handful of animals.

In a series of dives during the spring of 1979 the deep submersible *Alvin* hit the jackpot in its search for ore-forming eruptions on the sea floor. Evidence had been found that metal-rich layers are being laid down by eruptions of hot water along the East Pacific Rise and the central rift valley of the Red Sea. The island of Cyprus bears witness to the fact that some of the world's most valuable mineral deposits were formed in this way. Most of the early Mediterranean civilizations—Phoenecian, Egyptian, Greek and Roman—derived their copper from Cyprus and the Romans called the metal cyprum (later modified to cuprum from which "cu," the chemical symbol for copper is derived). The metal deposits on Cyprus, as in many other regions, are in the form of sulfur compounds (sulfides).

Along the Galapagos Rift Zone, west of Ecuador, sea floor explorers aboard the *Alvin,* operated by the Woods Hole Oceanographic Insti-

tution in Massachusetts, had already found evidence of hot spring activity and colonies of specialized deep-sea creatures dependent on the eruptions, such as giant clams and worms living inside ten-foot tubes of their own making. They had not, however, witnessed anything to compare with the tall stacks emitting jets of metal-blackened water that they saw on the East Pacific Rise south of the southern tip of Baja, California. The East Pacific Rise is a gently sloping ridge that bisects the South Pacific, paralleling the coast of South America. It is a "spreading center" on either side of which the sea floor pulls away, opening the oceanic crust and allowing molten rock to rise into the gap. Sea water percolating through this newly erupted rock becomes greatly heated and leaches various minerals from it. The geysering water observed from *Alvin* was at about 700 degrees Fahrenheit, erupting from stacks six to fifteen feet high. As soon as the erupted water encountered the frigid oceanic bottom water, only a few degrees above freezing temperature, the sulfides of copper, iron and zinc in the geyser plume precipitated and fell to the sea floor, forming mounds 50 or 60 feet high. As one scientist aboard *Alvin* said later of the spectacle, "it was like Pittsburgh in 1925 with all those blast furnaces going full tilt."

South of the Galapagos Rift Zone the drill ship *Glomar Challenger,* under two miles of water, sank a hole 1,100 feet into sea floor that had presumably been formed along the rift (itself a spreading center) some five and a half million years earlier. The ship had drilled 800 holes at 501 sites throughout the world's oceans and, while primarily financed by the United States National Science Foundation, was now an international enterprise, with million-dollar-a-year contributions from Britain, France, Japan, West Germany and the Soviet Union. This hole, however, was special in that for the first time it was used for an extended period as a laboratory within which a variety of experiments could be conducted including efforts to sample water that had percolated through the deep rock, imaging of the hole's walls by a special sonic device and the lowering of a Soviet magnetometer the full length of the hole. By sampling the sediment layers laid down on various parts of the sea floor as well as the bedrock beneath them the *Glomar Challenger* has probably added more than any other vessel to what is known of the history of the oceans and their inhabitants.

The Rising Tide of Oil Spills

by Michael Harwood

Observe the broken ship—a large object lesson. The Amoco Cadiz. When she was still in one piece, she was as long as three football fields. A few miles from the Brittany coast of France the stormy night of March 16, 1978 she lost all her steering; her captain called for help, a tug came to her aid but did little good instead. French officials have charged that the tug and tanker captains wasted time arguing over division of the salvage. After a few hours, the huge and helpless supertanker was blown on a reef near Portsall, north of Brest, and there began to leak her oil. For days, hope was held out that much of the 220,000 ton oil cargo could be pumped off into other vessels. But bad weather and high seas prevented that, and, meanwhile, the Amoco Cadiz was spilling her oil. It became the worst spill in history.

This is the Age of Oil Spills. Torrey Canyon, Argo Merchant, Amoco Cadiz—on and on and on. A tanker gets in trouble—through bad judgment, bad equipment, bad luck—runs aground near the beach and starts spilling oil. In the process fishing grounds, shellfish beds, lobster pounds, seaweed harvests may be seriously damaged; a year's tourist trade may be destroyed. With nearly 4,000 ships carrying some 11 billion barrels of oil each year, such disasters will happen again and again, on all the coasts of the world. That is a mathematical certainty.

If we can't stop all the oil spills, can we at least clean up the messes they cause? That takes technology and trained people, and cleaning up oil spills is a new and primitive art. Equipment and techniques do exist—floating booms and chemicals to corral the oil, pumps and skimmers to pick it up, straw and "sorbents" to mop it, chemicals to disperse it, materials to sink it, microorganisms to eat it. But even the most commonly used of these techniques are still quite limited. And most of the chasers of oil spills, worldwide, are only just learning how to muster people and equipment effectively when a major spill occurs.

If a spill threatens in a quiet, sheltered place, in calm weather, and it doesn't involve massive amounts of oil, we have the techniques and equipment and experience, in many harbors of the world, to pump the oil out of the stricken ship and into another vessel. If the spill actually occurs, we can get a lot of the oil out of the water. But if the accident happens outside that harbor, in rougher

A recent calamity in the Age of the Oil Spill: On a stormy night, the supertanker Amoco Cadiz lost its steering, foundered and, shortly thereafter, began to leak its 220,000 tons of heavy crude oil.

weather, and the ship is carrying millions of gallons of oil, our ability to stop the oil from causing serious harm is drastically reduced.

Most people in the oil-spill cleanup business will readily admit the limitations on what they can do—that their chances for success generally improve the closer the oil gets to shore. In fact, cleaner-uppers do best wherever they can get a lot of careful human labor working on the spill, cleaning it up inch by inch.

"That was what was so costly about the St. Lawrence River spill in June 1976," said Charles C. Bates, science adviser to the commandant of the United States Coast Guard. "The cleanup was done the same way as in China—little girls carrying rags, wiping rocks, or with a putty knife, scraping the gunk off." That appeared to be one of the best methods available to the French in the spring of 1978 on the oil-soaked shores of Brittany.

The problem of oil-spill cleanup would not be so considerable if petroleum always had the same chemistry and consistency, and if it always happened to spill in indoor swimming pools, nowhere else. In fact, however, even the crude oils— petroleum as it comes out of the ground—vary

widely, depending on where they are produced; and the refining process that makes both fuels for machines and basic materials for plastics and medicines adds almost infinitely to this variety. At one end of the spectrum, petroleum is gasoline— very poisonous at full strength to aquatic creatures, very volatile and light and quickly evaporated. At the other end is refining's "last cut," Number 6 residual oil, a heavy slow-burner used in industry, the stuff dumped by the Argo Merchant in 1976.

The potpourri of oils goes into the water in all kinds of circumstances. The water may be salty, fresh, shallow, deep, warm, cold, sheltered by surrounding terrain, exposed to the weather, subject to tides, pulled down by gravity; the water may flood a salt marsh or a freshwater swamp; the bottom and the margins may be mud, sand, gravel, boulders; and the spill may occur in weather conditions ranging from a flat calm in April to a hurricane in September to a blizzard in January. It would be no mean trick—in fact, even to contemplate it might be considered hubris—for man to develop a technology that would cope with all these possible circumstances.

Workers laboriously sweep the oil from a Brittany beach; such work is still necessary, despite new cleanup techniques.

No one worried much about oil spills at sea before the Torrey Canyon came a cropper in 1967, sending oil ashore on the Cornwall and Brittany beaches. So the cleanup technology was rudimentary: on shore, a mop-up with straw and pitchforks and farmers' pumps. Offshore: bombs to set the ship's oil afire and strong chemicals to disperse the oil in the sea. After the 1969 Santa Barbara oil spill was cleaned up with similarly crude techniques, including the use of 3,000 bales of straw, one oil-industry man remarked, "In an age when we can reach the moon, we should be able to do better than this."

Nine years later, we worry about it a lot more, and we *are* able to do better.

● *Containing the oil.* First of all, we can better contain oil once it leaks out of a ship. A spill usually spreads out on the surface (though oil has an affinity for itself, and how far it will spread depends on the kind of oil involved). The wind or a current or both usually moves it as well. Since most oil does float, a floating barrier—a boom or string of booms—can be set to capture or intercept the oil. When the moving oil hits this obstruction, it may stop moving long enough to be mopped or sucked from the water. Booms have been used for years in urban harbors, where oil spills are more or less chronic.

However, these floating booms have severe limitations. In even a very gentle current, the efficiency of the booms drops off, and if the water under the oil is moving much faster than one mile an hour past the booms, virtually all the oil will go with it. "Entrainment" is the reason—the "collection and transport of one fluid by another moving at high velocity." When the flowing water hits the boom, it stirs up the water behind it and ducks under the boom; the stirring action breaks up the oil into "pieces," which are sucked under the boom with the water.

The best way to use booms in a current, in fact, seems to be not to try to block the oil and hold it in one spot. Instead, the booms are allowed to move with the flow of the water, with just enough drag to let the oil catch up to the booms but not so slowly that the moving oil escapes under the boom. The booms can also be angled into the current rather than across it, to deflect oil into quiet water; that increases the usefulness of booms in fast-moving currents. Special current deflectors are also being

27

developed to steer oil out of the main stream of a river.

A new technique involves using chemicals to "herd" the oil. These herders have the effect—again if the water is not very rough and the current is not very fast—of encouraging the oil's natural tendency to stick to itself: Surrounded by herder chemicals, the oil rolls itself up into a compact mass, comparatively convenient for mechanical pickup.

● *Picking up the oil.* Oil can be lifted out of the water with various kinds of collectors and pumps and hoses, then separated from any water that may come with it, stored someplace for transport, and eventually consumed as originally intended. Vacuum trucks, like those that clean out septic tanks, are often used for sucking and carrying oil or oily water that has come close to dry land.

"Skimmer" equipment that mechanically picks up oil on moving belts or on rotating drums or plates is sometimes fixed permanently in place—bolted to a dock, say—to recover chronic oil spills. And a good many skimming devices now in use are actually self-propelled vessels with wide-open mouths for swallowing oil.

The French have developed skimmer gear that simply scoops up oily water and separates it into oil and water (it's said they had difficulty with this gear during the Amoco Cadiz spill). The French skimmer has one important advantage: It can be easily and quickly attached to most boats.

After only nine years of development, this skimming equipment—most of it small and of low capacity—has proved itself moderately effective in sheltered waters. But even the best of the skimmers are not much use when the waves are higher than four or five feet. When a skimmer is rocking in high seas, it cannot maintain a continuous intake of oil. The wave action breaks up the oil, mixes it in the water and shoves a lot of it below the skimmer's reach.

There are other pickup methods being worked on that, like straw, the old standby, literally mop the oil from the water. Straw itself is still used on occasion; for example, a double fence of chicken wire stretched across the bed of a tidal creek and stuffed with straw catches the oil sloshing back and forth with the tide. Various manufacturers have developed more effective materials, called sorbents, which collect oil and reject water. Pads or chunks of sorbents can be scattered directly on a spill; however, the next step in that process is often

no different from the next step with straw. The pads must be recovered manually, for burning or burying.

There are now skimmers that eliminate this final step; they use sorbents in endless oil-capturing belts that loop through large pulleys anchored in the spill. The sorbent belts are squeezed out automatically and fed back into the water.

● *Making it go away.* One way of making oil "disappear" is to mix something with it that makes it sink. The French used some chalk to do this after the Amoco Cadiz spill in March 1978. The drawback to such a method is that you are simply moving the hazard to marine life from the surface down to the ocean floor.

The most controversial method for treating oil spills is to disperse them chemically. The idea is to break the oil up into small drops that can then undergo natural biodegradation and evaporation. Since oil tends to stick together in rather large patches or streaks on water, chemical dispersants are used to change the surface tension of the oil, to break it into droplets that will not readily recombine. The dispersed oil will also drift away into the whole water column, top to bottom.

Dispersants can make an oil spill "go away" or seem to disappear before it causes any serious visible damage to beaches or birds. This makes dispersants extremely attractive to shippers and oil companies, because, for them, oil spills are, as much as anything else, political, bad-publicity events that make other people mad at them.

The first dispersants—such as those used on the beaches of Cornwall when the Torrey Canyon's oil came ashore—were very toxic. Since oil, too, is toxic, this double dose hurt a lot of sea life.

The British have continued to use dispersants as less toxic formulas became available, because they believe (with justice) that mechanical means are simply not up to the job.

But harm done by the original dispersants helped discourage some nations—including the United States and Canada—from using the chemicals. This country's regulations allow dispersants only for crises of the first rank—when there is risk of fire or the endangerment of human life, the immediate threat to a large wildlife population, a fish spawning ground or a shoreline where the arrival of oil would have absolutely unacceptable environmental, economic or political consequences.

The Amoco Cadiz took a heavy toll of wildlife, including this oil-drenched sea bird. Oyster beds were also badly damaged.

Officials hesitate to use even low-toxicity dispersants because they know that oil, when left alone, will eventually dissipate and break up into progressively smaller clumps in a natural rhythm set by the weather and the water action, releasing its poison fairly slowly. The cleaner-upper who breaks up a spill rapidly with chemicals may smack the area with a terrific—if short-lived—belt of toxicity.

So the argument now comes down to this: If you are going to do anything about the oil spill, should you add something else to the water that doesn't belong there, or should you try to get the oil out?

● *Why not burn the oil?* Burning as a cleanup method has often been suggested—by a generation that came of age watching World War II films in which the submarine blew up the tanker and covered the sea with floating, flaming oil. Fires have been used to clean up badly-oiled marshes. But on water—generally speaking—oil does not burn well—and for a very simple reason. Even if you manage to set it afire, it won't stay hot enough to burn, because cold water is constantly chilling it, as well as splashing on the fire.

The more easily combustible parts of the oil may burn, but even then, the oil must be ignited somehow and encouraged to keep going. Various "wicking agents" are commercially available, but their reliability and practicality have yet to be demonstrated under crisis conditions at sea.

The Swedish Coast Guard has used fire to clean up oil spills trapped in ice, but that technique is rarely called for.

● *Biological cleanup.* One of the most intriguing new methods involves oil-eating microorganisms. Water environments, both fresh and salt, are accustomed to oil. Natural seeps on the bottom have been pumping oil slowly into water for millenniums, and microorganisms that like to eat oil gradually evolved. In an effort to take advantage of that, a commercial product—a mix of animals—has been developed for use in salt water, and a companion is now being developed for fresh-water application.

But the oil eaters have their limitations. They find some kinds of oil easier to digest than others. And they need other chemicals on the plate—oxygen, nitrogen, phosphorus—in order to get

29

the oil down, chemicals that may not happen to be available in sufficient supply at the site of a spill. On the whole, this is being treated as an interesting avenue for future work; it is not considered an important cleanup technique now.

□

So, some progress in cleanup technology is being made and a good many firms here and abroad now produce cleanup equipment. But compared, say, to the sophisticated gear and techniques available for oil drilling and oil pumping, this is still a tinkerer's and handcrafter's technology. Looking at the equipment and listening to the stories of its development and sales records, one is reminded of the auto industry in those colorful days before the Model T.

To be sure, the equipment is being bought in the United States, for example, by the Coast Guard, the Navy, some 50 private cleanup contractors and about 100 local cleanup cooperatives organized by the petroleum industry. The Environmental Protection Agency and the Coast Guard and the American Petroleum Institute jointly sponsor an oil-spill conference every other year to discuss the problem and hear papers on new developments and to display the latest equipment and materials. Nearly 1,700 people, representing 26 countries, attended a three-day meeting in New Orleans in March 1977. But despite such interest and effort, there is a great deal that the technology and its users can't manage yet.

First, there are organizational problems. It's one thing to have equipment commercially available and techniques already developed for cleaning up spilled oil. It's another to have enough of the equipment within reach when it's needed and to be well enough organized so that trained people and proper equipment can get to the scene of a spill quickly to start cleaning it up.

Dr. Jerome Milgram, professor of ocean engineering at M.I.T., has noted that a total spill system includes booms, skimmers, storage vessels for the oil, tow vessels and trained personnel. "If any one of these elements is absent," Dr. Milgram said, "even if all the remaining items are provided, essentially no oil cleanup can take place."

In fact, a lot of the things called for by an emergency—ships, helicopters, pumps, trucks, bulldozers—are frequently tied up in other jobs,

and then, of course, much of the equipment has to be modified before it can be used.

□

Despite this, North American responses to oil spills are pretty well organized, comparatively speaking, and they can get off the mark quite quickly. A National Oil and Hazardous Substances Pollution Contingency Plan has been in existence since 1970. According to Kenneth E. Biglane, director of E.P.A.'s Division of Oil and Special Materials Control, "In this country, we have 10,000 oil spills a year; 80 percent of these are less than 100 gallons. On top of that, we have the national contingency plan, and because we have so many spills—albeit small ones—we implement that plan almost daily in this country. That means our plan is sharp and ready to go."

The Coast Guard, as lead agency under the plan, heads a National Response Team, which works together all the time. Everyone knows who's supposed to be in charge and where, when trouble arises. They know where to hire tugs and barges and other equipment because it's been done so often before.

"It's like fire-evacuation plans in a skyscraper," says Lieut. Michael Donohoe of the Coast Guard's Gulf Coast Strike Team. "They've got them to comply with the regulations. But unless you have fire drills or evacuation drills, if the day ever comes to actually get out of that building, there may be a hell of a problem."

Just so. The French had a response plan, but they had not had enough practice to cope with an accident the size of the Amoco Cadiz spill.

The difficulties being encountered in cleaning up oil spills on water are certainly not unusual, considering the complexity of the situation and the short time the industrial nations have been interested in reacting to it. On top of that, the required sense of urgency about seeking certain levels of solutions to the problem is weakened somewhat by arguments and indecision about related matters. The best method of containing oil on water, after all, is to keep the stuff in the tankers, so ship design and navigation equipment and personnel training and emergency procedures and similar factors for which higher standards might be set are now being debated—often heatedly—and these involve complex international maritime relationships.

*French soldiers help local farmers clean up the beach polluted by oil
spilled from the wrecked Amoco Cadiz.*

The question of liability is also being argued, nationally and internationally: Who should be physically responsible for cleaning up a spill, and who should pay for the damage? Some oil firms (such as, apparently, the Amoco International Oil Company in the Amoco Cadiz case) will accept a great deal of responsibility, but others have ducked it.

Furthermore, it is useful to put these highly visible oil-spill accidents in perspective. Only a small fraction of the oil in our water—perhaps 3 to 6 percent—gets there from the photogenic sort of accidents you see in the newspapers. Some oil, maybe as much as 10 percent, leaks into water naturally, from seeps in the ocean bottom.

And possibly as much as *85 percent* of the oil in our planet's waters comes from the undramatic drip-drip-dripping of our tail pipes and bilges and sewer connections. Tanker captains spill oil on purpose when they clean their cargo tanks and bilges, although industry and governments are trying to halt the practice. The rivers of the world swallow a great deal of oil on a daily basis; it comes from industrial wastes, leaking tank farms, street runoff and city sewers. Much of it comes from a most prosaic place — our automobile crankcases — used oil poured into storm drains and sewers by auto and truck oil changers — as though the drains and sewers did not lead into our lakes and rivers and oceans. Other sources for oil in water include daily offshore production, coastal refineries and atmospheric rainout.

These ratios and percentages make governments hesitate to commit themselves to large expenditures for standby cleanup equipment, however much the public becomes excited when a big oil spill occurs near shore.

And no one really knows just how serious the long-range environmental effects of a big oil spill *are*. We don't know because we haven't been studying them for very long. United States law says that putting any oil at all in navigable waters is illegal. The product is poisonous and otherwise hazardous, and from an environmentalist's point of view, that's a good base line.

But the practical answer to the question does not appear to be so black and white. Richard Dewling, a regional director of technical programs for E.P.A., points out, for example, that "an oil spill of 50 gallons in a shellfish bed is very critical. A spill of 1,000 gallons in, say, a Bayonne shipyard is not that critical. Sure it's *critical*, but the response time, the concern you must have, your response mode, cleanup method, must be completely different." And then Dewling goes on to try to add greater perspective to the question, by talking about "political oil spills."

"There are oil spills that are of severe potential environmental consequence," he says, "and there are oil spills that politically and socially are very sensitive.... If we had a spill on the July 4 weekend, off the coast of Atlantic City, we would take one approach to cleaning up; if we had it in December, we'd take another." Public outcry and the threat of public outcry have had a crucial impact on the way government and industry have approached the problem of oil spills; and a good many of those closely involved believe the impact has in some respects been deleterious.

Sometimes the environmentally best thing to do, they suggest, may be just to allow the oil spill to break up on its own, so it can be slowly metabolized into water and carbon dioxide. Dispersants may do a lot of immediate damage; cleanup crews and bulldozers and vacuum trucks may harm a shoreline or a marsh far more than the oil itself will. But currently, as Richard Dewling says, "the public would not allow you to make that type of decision, because the press will kill us. They'll say, 'There are these dumbbells sitting there with their fingers in their mouths, doing nothing, watching the oil go back and forth.'"

That was just the sort of thing being said about the French Government in the first weeks after the Amoco Cadiz went aground. True enough, the French were not ready to put their spill-response plan into operation quickly. And quite likely the Government was distracted by bureaucratic infighting and by the national election that coincided with the spill.

But American observers returning from the scene were quick to point out that until you had seen for yourself what was happening on the Côte du Nord, you could have no conception of the size and physical impact of the spill. "It is staggering," said Ken Biglane of E.P.A. "Not too many spills have impressed me since the Torrey Canyon. But this one surpasses anything in my wildest envisionment of an oil spill."

Quite apart from the problem of containing the oil and mopping it up, where do you put 68 *million* gallons of crude oil? That's roughly 1.36 million

Big waves kept rolling over the wrecked supertanker and its oil continued to pour out.

50-gallon drums worth—doubtless enough to line a route between Portsall and Paris several times over, if you care for that sort of mathematics. There's very little coastal oil barge traffic on the Côte du Nord, for instance, so barges could not be called out to carry the oil. The logistics of putting together a fleet of tank trucks to pick up the oil at the water's edge were extremely complicated.

In the interim, the French dug ditches above the high-tide line, lining the ditches with plastic, and pumped and shoveled and scraped the oil into those ditches. It might look foolish and even hopeless to the distant observer, but it was the best that could be done under the circumstances.

Returning American experts did not hesitate to point out that however well equipped and staffed and prepared for oil-spill cleanup a nation might be today, a spill the size of the Amoco Cadiz spill, held against the shore by high winds, would do much physical and ecological damage before it could be cleaned up. That's a sobering thought for coastal communities everywhere. The world is only beginning to grasp the enormity of the day-to-day risks that are part and parcel of the Age of the Supertanker, the Age of the Oil Spill.

This is Not the End of the Game

by Boyce Rensberger

Nairobi, Kenya.

The teeming herds of wildebeests and zebras are now nearly gone. The browsing families of elephants no longer wander across the plains. The rhinos that once pawed the dust of almost every valley are hard to find. The lions and cheetahs are absent from vast savannas that have been turned into cattle ranches and wheat farms.

The hippos and crocodiles that once crowded the riverbanks are scarce. The gazelles, impalas, kudus, oryxes, gerenuks, hartebeests, dik-diks and other antelopes—all these are dying out. At the same time, Nairobi's curio shops are now stacked with zebra hides, elephant tusks, rhino horns and assorted other trophies.

Those who knew Kenya a generation ago can still remember vast grasslands where one could ride a horse for hours through endless herds of antelopes and zebras. Professional hunters can recall a time when no one talked about the end of the game. Ivory merchants, who operated quite as boldly under British colonial governors as they do now in independent Kenya, believed there would always be tens of thousands of elephants.

Westerners thought that there would always be a "someday" when one could come to Kenya for an incomparably exciting experience—whether sighting down rifle or camera lens.

Yet, a land that once seemed inexhaustible in its abundance is now reaching exhaustion. A land that once seemed unconquerable in its pristine wildness, is now virtually conquered.

There are those who believe that we have come to a point of no return in East Africa, that we have presided over an ecological disaster of historic proportions. They blame the white man and the legacy of colonial exploitation, they blame corrupt black officials, and they say that the whole world is the loser. There are others who feel that the agricultural and economic needs of Africa's population must take precedence over the needs of wildlife. And there are those who say that wildlife can be made economically attractive enough to permit the coexistence of animals and people.

There is no doubt that the glory days of Africa's wildlife are now over. But the story of wildlife in East Africa is not as simple as it might seem. It is not only a tale of "the earth's priceless natural

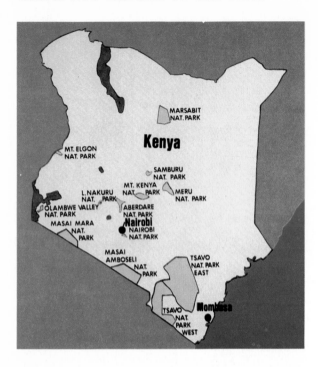

heritage" being plundered by greedy poachers. Far more significant in the long run is the impact of burgeoning human populations that have little choice but to encroach on wild habitats.

The existence of both rampant poaching for profit—often protected by corrupt government officials—and legitimate growth and development, is forcing conservationists into some of the most difficult choices their movement has had to face. Herds of elephants, for example, are being squeezed into smaller and smaller territories; the resultant densities are destroying the vegetation that beasts need to survive. Thus, many conservationists have been forced to advocate shooting some of the elephants to bring the numbers back into balance, even as they campaign against the illegal shooting of elephants. In some cases the seemingly paradoxical intentions involve the same elephants.

The intensifying competition between man and beast in East Africa has raised a profound challenge to the motives for wildlife conservation. As the conservationists' battle to protect dwindling numbers of wild animals intensifies, the traditional arguments for conserving animals, which used to go unchallenged, are now subject to examination with increasing pragmatism and even skepticism.

How does one balance an American's desire to know that there are elephants or lions or giraffes living free in the world, against an African's desire for enough land to feed his family? How many wild elephants does it take to meet our wish that there be wild elephants to think about, or to go and see someday? Would the extinction of the cheetah really trigger an ecological disaster, or hardly a ripple?

Ernest Hemingway or Isak Dinesen or Teddy Roosevelt would be saddened to see the wildlife in Kenya today. It is nothing like what they knew. But if you have never been to Africa, no other wildlife experience can prepare you for the spectacle still to be seen there. And there is every likelihood that a good share of this will always be there, at least in a few heavily guarded regions.

□

A thorough examination of the state of wildlife survival in Kenya demands not only a knowledge of wild animals but also an understanding of the legitimate needs of human beings living on the same land. And, of course, it helps to keep in mind that as long as there are people willing to pay handsomely for ivory, horn and hide, there will be others quite prepared to supply them, no matter what the law. Such markets do exist in abundance, overwhelmingly in the world's richest countries.

Although human encroachment on Africa's animal habitats poses the most serious danger to wildlife, it is the struggle between the poachers and the antipoachers that constitutes the most controversial aspect of wildlife conservation in Kenya today. Depending on whom one talks to, the situation is either hurtling toward widespread extermination of wildlife or showing encouraging signs of coming under control. Even the most pessimistic conservationists, however, see at least some basis for hope, and believe that if protection can be instituted in time, wildlife numbers will eventually build up again.

Amboseli National Park, 150-square-miles of bush and swamp that has long been Kenya's top money-making wildlife area, illustrates both the seriousness of the wildlife decline and the hope of improvement.

Until a few years ago, Amboseli was one of the best places to see rhinos. There were some 60 or 70 in the areas frequented by tourists. But the Maasai, the proud nomadic cattle herders whose land the park once was, have been spearing them both as proofs of manhood and in retaliation against the

In Tsavo National Park, there were an estimated 36,000 elephants in 1974.
In 1977, sample aerial counts produced an estimate of around 20,000.

government for moving them out of their traditional grazing lands. Tourists used to be able to see 20 or 30 rhinos in a day. Now they usually see none. There are only one or two rhinos left.

But the situation may be improving. The Kenya Government recently began making cash payments directly to the Maasai in compensation for their lost lands, and, according to Dr. David Western, a University of Nairobi zoologist studying Amboseli and the Maasai, Maasai attitudes are changing.

Not long ago, for example, a Maasai man ran eight miles to report that poachers had killed a rhino in a remote area of the park. Rangers were immediately dispatched and the poachers were caught with their trophy, a single horn that on the black market would have brought $650—more than the average Kenyan laborer's annual wage.

While the poaching around Amboseli and other wildlife areas in southern Kenya is mostly done by individuals for money or meat, a far more dangerous form of poaching is rampant in Kenya's sparsely populated northern regions.

"In the north there are armed gangs using very sophisticated weapons, machine guns and trucks," said Menasses Otto Keiller, who directs the Kenya Government's 75-man Anti-Poaching Unit. Keiller, a Kenyan despite his European-sounding name, is a powerfully built ex-army officer who loves animals and regards wildlife as one of his country's greatest assets.

"Up here we're not dealing with the little man trying to feed his family," Mr. Keiller said. "It's much worse. We get bands of 10 or 20 armed men and they shoot all the game they can find. They're protected by 'big shots' in Nairobi."

The Anti-Poaching Unit, which must cover all of Kenya's 225,000 square miles, is based just outside

For some time, conservationists have warned that wildlife is disappearing from Africa's hills and valleys. Now one rarely sees elephants outside the national parks.

Isiolo, a dusty little town in central Kenya between the heavily populated southern half of the country and the vast, semi-arid bush country to the north. The northern region is inhabited by growing, nomadic families of cattle, goat or camel herders and declining populations of elephants, lions, antelopes, zebras and other game. "I'll show you something," Mr. Keiller said to a visitor as he opened the heavy steel door between his office and a room where he keeps seized trophies.

A dozen or so tusks were stacked at one end with a few stiff leopard hides. Sad enough, but nothing to compare with a bulging gunnysack containing some 30,000 finger-length horns from the tiny antelope species known as dik-dik. The horns were destined for the booming Nairobi curio trade, where they would be mounted and sold to tourists as jewelry or key rings.

It is the curio shops that provide the most visible evidence that all is not right with Kenya's wildlife.

There are scores of them, most stuffed with elephant tusks, zebra hides, mounted antelope heads, lion-claw necklaces, monkey-skin throw rugs, ostrich-foot lamps, gazelle-hoof key rings and assorted other animal products eagerly snapped up by wealthy tourists, principally Americans and Europeans.

The number of such shops has grown dramatically over the last five years, the same period in which legal hunting was restricted and, in 1977, banned entirely. If there is no legal hunting, where do the animals come from?

One legitimate source of at least a few animals is the Government's game control program which, quite properly, shoots the few elephants, lions and other animals that menace villages and destroy crops. Also some of the ivory comes from elephants that die of natural causes.

These sources, by all reliable accounts, could not supply the quantities to be seen in the shops.

Furthermore, wildlife experts note, it is seldom necessary to "control" minimally destructive species such as zebras and antelopes.

Curio dealers, when asked about their sources, uniformly reply that their supplies are all legally obtained. "We get it all from the Game Department," one dealer said. "We don't buy from anybody else. We used to buy from hunters, but that's finished with the ban."

In this man's shop, typical of the larger ones in Nairobi, there were, among other things, more than a hundred uncarved elephant tusks, scores of carved ones, a dozen lion-skin rugs with heads set perpetually asnarl, and several skins from the endangered, tree-dwelling, colobus monkey. Another shop in Nairobi had 35 skins from the rare Grévy's zebra, a species with narrow stripes that, by some accounts, numbers no more than 2,000 in Kenya. The skins sell for prices approaching $1,000 apiece.

Reminded that Kenyan wildlife was rapidly declining, one dealer replied, "Yet, it is so. Soon it will all be finished. The game is going. Next year the prices will be even higher. Would you like to buy?"

Even Kenya's Ministry of Tourism and Wildlife, which regulates all wildlife conservation and trading, concedes that at least some curio shops deal in illicit trophies. "Curio shops should get their ivory from Government stocks, but they don't always," said Yuda Kamora, the ministry's permanent secretary. Mr. Kamora said the Government has raided some curio shops and prosecuted the owners for lacking official permits for their ivory. "We are doing the best we can to crack down but, like all governments, we have limited powers and resources, and some thieves and smugglers get away," Mr. Kamora said.

Although Mr. Keiller's 30,000 dik-dik horns have not yet reached the curio market, they may yet get there. Mr. Keiller had been holding them as material evidence for a trial of the poachers but, without explanation, the trial has been put off. Just recently an order came through from the Ministry of Tourism and Wildlife to release the horns for sale to the curio merchants.

It is not the first time such orders have come down. On another occasion Mr. Keiller's men arrested a man in a Mercedes carrying 30 tusks. The response from higher up was an order to release the man, and a reprimand for Mr. Keiller.

Even when poaching cases do come to trial, the administration of justice is often uneven. One edi-

Amboseli National Park is on land that once belonged to and was used for cattle grazing by the Masai.

tion of a recent Kenya newspaper, for example, carried two items about poaching trials. In one case, a woman belonging to the politically dominant Kikuyu tribe was charged with possession of 50 elephant tusks. Charges were dismissed. In the other case, a member of the smaller, almost power-

Today the Kenyan Government is protecting its wildlife in a variety of ways.

less Giryama tribe was sent to prison for possessing three giraffe tails.

Dozens of similar stories — vigilant antipoaching efforts repeatedly thwarted by orders from various corrupt high government officials — have circulated in Kenya for years, to the growing disillusionment of those working to protect game animals.

Tony Dyer, president for 14 years of the East African Professional Hunters Association and a well-known figure to responsible sport hunters all over the world, is one of the few who will put his name to such accusations.

"To my certain knowledge," he said, "there are four officials using their positions to obtain income from illicit game trophies." Mr. Dyer, who disbanded his organization a few months ago when Kenya outlawed sport hunting, said the four held the ranks of senior administrative officers, members of Parliament or assistant ministers. He made it clear that there was plenty of evidence that a number of others were also involved.

Mr. Dyer cited one instance in which a truckload of ivory was captured by the Kenya police but released within the day when an assistant foreign minister showed up in person to order charges dropped.

Mr. Dyer believes the main reason the Government banned sport hunting was not to eliminate the relatively small toll that hunters took, but "to keep our prying eyes out of the country." Mr. Dyer contends that the professional hunters who were required by law to guide visiting sport shooters, also served as antipoaching sentries. "We were the only ones in the bush who cared about the game," he said. "Without us there, there's nobody but the poachers."

Mr. Dyer is fiercely proud of the achievements of professional hunters, who not only kept trigger-happy clients from killing themselves or needlessly maiming wildlife, but who also won legal protection for cheetahs, lionesses and female rhinos.

Mr. Dyer and a small clique of British settlers and their descendants who came to regard this strikingly beautiful land as a precious thing, have fought for years to create and defend national parks and reserves. They stayed when Kenya won her independence in 1963, and have worked closely with emerging African leaders to perpetuate and improve conservation programs.

Through its own efforts and with the assistance of men like Mr. Dyer and of international conservation groups, Kenya has done, on the whole, a reasonable job of protecting wildlife, often spending a larger share of its budgets on conservation than any wealthier country.

"There will never be, in the history of this world, finer game fields or a finer philosophy for the management of those game fields than we had here in Kenya," Mr. Dyer said. "What we fought to achieve has been completely undone by John Mutinda."

When conservationists, park wardens and others speak of the corrupt Government officials who protect the poachers and profit from the curio trade, they seldom mention names. But when they do, John K. Mutinda is usually one of them. Mr. Mutinda is head of the Wildlife Conservation and Management Division of the Ministry of Tourism and Wildlife. He is in charge of the parks, the antipoaching program and the game "control" program.

It is safe to say that nearly everyone in a position to know Kenyan wildlife matters behind the scenes believes Mr. Mutinda to be heavily involved in the protection of poachers. They note that it would be virtually impossible for many questionable activities to take place repeatedly over the years without at least his acquiescence. However, no formal charges or documentary evidence have been put forth to substantiate the allegations about Mr. Mutinda, and efforts to obtain an interview with Mr. Mutinda were unavailing.

The other name mentioned most prominently is that of Kenyatta. Rumors have circulated for years that relatives of President Jomo Kenyatta—chiefly his wife, Mama Ngina Kenyatta, and his daughter, Margaret Kenyatta—have been dealing in ivory of questionable origin, exporting hundreds of tons of it to buyers in Hong Kong and elsewhere. Documentary evidence confirming these ivory exports has been published in the British and American press.

The Government has never disputed that the Kenyattas dealt in ivory, but it has asserted that the dealings were legal since they were carried out under licenses issued before private ivory exporting was banned in 1973. Official Government sales of "found" ivory, and ivory confiscated from poachers, are still permitted. The legal status of the ban has long been in dispute. The Government now contends that it did not take effect until last year, when a formal notice was published in the Kenyan equivalent of the Federal Register.

There have been reports that the Kenyattas have quit the ivory business to deal in coffee, now a highly lucrative commodity. Mama Ngina Kenyatta is said to be entirely out of the ivory business, although Margaret Kenyatta remains chairman of the United Africa Corporation, and exporter of many commodities that, at least in the past, have included ivory.

Miss Kenyatta is Kenya's representative to the United Nations Environment Program, which is based in Nairobi, and reportedly has been under pressure from international environmental leaders to get out of the ivory business before Kenya's reputation for competent government is irreparably tarnished. British and American press accounts over the last two years of the Kenyatta family's ivory dealings have severely embarrassed members of the Kenya Government.

In 1976, in a rare display of displeasure toward a cabinet minister, Kenya's Parliament voted almost unanimously to investigate the Ministry of Tourism and Wildlife. Amid strong but undocumented charges of corruption, and of the ministry's masterminding poaching rackets, the motion carried with only one dissent, that of the minister himself, Mathews Ogutu.

Mr. Ogutu has also been the target of a long editorial campaign by Kenya's leading daily, The Nation, which charged him with failing to curb the decline of Kenya's wildlife. In a rare, front-page editorial, The Nation called for his resignation.

Many people in Kenya believe that the spring 1977 announcement of a total ban on sport hunting was an attempt by Mr. Ogutu to redeem himself in the eyes of conservationists. The move, in which Kenya cut off an estimated $1.4 million in income from hunting licenses and much more in related business income, was hailed by many conservationists. Their major reservation was that the action did not include a closing of the curio shops.

Mr. Keiller's antipoaching rangers, though hampered by a lack of funds, continue to seize major shipments of ivory and horn and other trophies. But, Mr. Keiller says, when the order comes down from Nairobi to release the trophies, "We get demoralized. The big guys sit in Nairobi. You can catch the little man doing the work but

The hippos and crocodiles that once crowded the river banks are scarce.

you can never get the big fish."

Even if all the big fish in Nairobi were put out of business, however, Mr. Keiller and his men would still face a sizable problem from Somali poachers. Somalia, Kenya's neighbor to the northeast, has long pressed territorial claims on a huge bordering region of Kenya, much as it has contested the Ogaden region of Ethiopia. As in Ethiopia, many of the residents of these disputed areas of Kenya are ethnic Somalis.

Mr. Keiller said some captured Somali poachers have told him they were sent into Kenya by their Government both to get to know the terrain and to shoot all the game they could, with the intent of provoking the Kenya government into a confrontation. Indeed, Somali poachers sometimes do not bother to collect the hides or tusks of the animals they kill. Others undoubtedly are selling the trophies to the curio trade or exporting them from Somalia to Europe and elsewhere.

No one knows exactly how many animals are killed by poachers each year. In fact, no one knows just how many animals Kenya or any other East African country has to begin with. The first complete national survey of wildlife in Kenya, financed by the Canadian Government, was begun in 1977.

Peter Jenkins, one of the few remaining white wardens in East Africa and one of the major figures in Kenyan conservation for a generation, estimates that in Meru National Park, which he administers, poachers are taking about 100

elephants a year out of a population of 2,500, and perhaps 20 rhinos a year from a population of 125. Even though the poachers are mostly using arrows and trying only to make a living, Mr. Jenkins lacks both the budget and the administrative powers that would enable his rangers to stop them.

"The time will come very soon when we'll have to deal with the organized poachers that are working outside the park," Mr. Jenkins said. "They've virtually eliminated rhino in the adjacent areas and, before long, these armed poachers will give this park their undivided attention. I hate to think of what will happen to Meru then."

In the vast Tsavo National Park, whose 8,300 square miles make it larger than New Jersey, there were an estimated 36,000 elephants in 1974, one of the largest such populations in the world. In 1977, sample aerial counts produced an estimate of around 20,000.

While the variations in such counts often are influenced by seasonal migrations, there is no doubt that several thousands Tsavo elephants have been slaughtered in the last three years.

David Sheldrick, another pioneer in Kenya conservation who served for many years as warden of Tsavo before his death in 1977, estimated that during a three-month period of 1976, there were nearly 1,400 poachers operating inside the park boundaries.

"Most poachers carry about five or six arrows, and they seldom leave the park until all these have been dischargd," Mr. Sheldrick wrote in a report to the Government just before his death. "The result of a slaughter on this scale must be a foregone conclusion. Dead and dying elephants are seen daily, and stocks of game cannot withstand the present onsaught for very much longer. Already, large areas have been completely denuded of wildlife."

□

Deliberate slaughter has not been the only threat to Tsavo elephants. In 1970 and 1971, thousands starved to death during a drought. Shorter droughts since then have killed more, bringing the total number of elephants killed by starvation in this decade to about 9,000. These elephants were killed not by poachers but by people who pushed the frontiers of agriculture and settlement into elephant country, driving the great beasts into the park's sanctuary.

Over the last 20 years, so many elephants have entered Tsavo that their normal rate of feeding on tree branches, and their tendency to push over or debark trees, has largely destroyed a once-dense vegetation. Tsavo, which used to be a lush land of bush and trees, is today a sad landscape of bare ground and patches of grass, strewn with the skeletons of downed trees and dead elephants. The rate of vegetation destruction is now so great that a normal dry season leaves little green food, so little that scores of other elephants, mostly babies and nursing mothers, die.

Biologists now fear that a continuing decline could trigger what they call a "population crash," wiping out elephants in the largest place set aside to protect them. A 1966 plan to study the elephant population and its reproductive potential by conducting autopsies on 3,000 elephants—which would have to be shot for this purpose—produced one of the bitterest controversies in East African conservation history. Many people, including Mr. Sheldrick, the warden, recoiled in horror at the thought of deliberately killing elephants in a park. The scientist in charge of the project, Richard Laws, perhaps the world's top expert on large-mammal population dynamics, was forced to leave Kenya. It was decided, instead, to "let nature take its course."

To date, that course has witnessed the starvation of three times as many elephants as the cropping program would have taken. And, while the cropped elephant meat would have been used for food,

carcasses of starved elephants are left to scavengers. And, because there is too much for the hyenas and vultures, most of the meat simply rots.

Although the poachers have, in a sense, cropped some elephants their own way, the numbers have still not fallen fast enough to catch up with the declining vegetation. In contrast to Kenya, Tanzania has adopted a policy of more active wildlife management. "The old method of 'letting nature take care of itself' becomes an abnegation of responsibility," said Derek Bryceson, a member of Tanzania's Parliament and director of its national park system, which includes the famous Serengeti National Park. Mr. Bryceson holds that as long as human settlements artificially constrict natural habitats, nature can follow only a distorted course unless there are wildlife management practices that compensate for the human impact.

These may include shooting some elephants where they are too numerous. Such a plan is being considered for Tanzania's Ruaha National Park and its adjacent Rungwa Game Reserve. Some 40,000 elephants there, more than the land can sustain, are destroying the vegetation and it is estimated that 5,000 to 10,000 may have to be shot.

By all accounts, poaching and government corruption in Tanzania are well under control, and that country's wildlife is surviving comparatively well. For one thing, Tanzania's population density is two-thirds that of Kenya, and its elephant numbers are greater, almost certainly running well over 200,000, and possibly growing. In contrast, Kenya is thought to have between 65,000 and 75,000, half the number estimated just three years ago.

For all the gloom that events in Kenya have fostered among wildlife enthusiasts, there is still cause for hope. Until coffee prices soared recently, tourism was Kenya's largest earner of foreign exchange. It is now in second place, and there is every sign the Government intends to maintain at least enough wildlife and wilderness to support this profitable industry indefinitely.

Unlike many countries, Kenya has a formally stated policy on wildlife. It says, "the Government's fundamental goal with respect to wildlife is to optimize the returns from other forms of land use. Returns include not only the economic gains from tourism and from consumptive uses of wildlife [sales of meat, skins and other trophies and, until recently, sport hunting], but also intangibles such as the esthetic, cultural and scientific

Those who knew Kenya a generation ago can still remember vast grasslands where one could see endless herds of zebras. Those pictured here are in Tsavo National Park.

The great numbers of elephants that naturally feed on tree branches and tend to push over or debark trees have destroyed a once dense vegetation.

gains from conservation of habitats and the fauna within them."

The policy, first set forth in 1975, goes on to outline the administrative machinery to carry it out and to explain, in pragmatic terms, how competing values for wildlife are to be adjudicated. The policy specifically rejects the goal of maintaining the largest possible number of animals. To have maintained that objective would have meant denying Kenya many opportunities for increasing its own food production. To ask that any developing country preserve all its wildlife is to ask that country not to try to feed itself. It is to ask the country to remain underdeveloped.

Kenya's new wildlife policy is in line with the emerging view of leaders in conservation that the surest way to protect wildlife is to ensure that it is economically justifiable. In other words, wildlife must pay its own way, either through earnings from tourism or from cropping on a sustained-yield basis.

The tourism argument has long been accepted. Cropping remains highly controversial, especially among the more sentimentally inclined. But support is growing.

Consider the case of cattle ranching on land that also supports antelope, zebra and other plains game. If there is no market for wildlife products, a sensible rancher would rather not have the game eating his cattle feed. He will not be averse to getting rid of the game. But, if the rancher can count on an income from game trophies or meat—either by shooting some himself, or by letting sport hunters pay to shoot—it is in his interest to limit the harvest to a sustainable yield so that there will always be game. Studies have shown that game used in this way is at least as profitable as cattle.

In a few large, private ranches in Kenya, measuring tens of thousands of acres, sport hunting of game coexisting with livestock once brought in appreciable sums of money. Some ranches, such as the Galana Ranch in eastern Kenya, even maintained private anti-poaching forces. With the hunting ban, however, this source of income has stopped, and the ranches can no longer afford the anti-poaching efforts. Photographic safaris earn much less than do hunting safaris.

The new Kenya policy also states, however, that where wildlife cannot earn more than some alternative land use, the animals will have to go. Mr. Kamora, the wildlife ministry's permanent secretary and a man widely considered a force for good in a ministry struggling with corruption, says the policy established three categories of land use. Some areas are designated as exclusively human territory (settlement and intensive cultivation), other areas are exclusively for wildlife (parks and reserves of which Kenya already has many and plans more) and a third category of mixed land use along the lines of the combination cattle and game ranch.

Even the most pessimistic of conservationists, such as Peter Jenkins, the Meru warden, believe this is a sensible policy. "Naturally, in certain areas wildlife will have to be subordinated to the needs of the people. This is to be accepted," he said. "I don't think irreparable damage has been done. Afforded proper protection in the wildlife areas, there's no reason the animal numbers shouldn't build up again."

Dr Iain Douglas-Hamilton, an elephant behavior specialist who is surveying the elephant populations of all Africa, also sees hope. "I'm not as pessimistic as some others are about Kenya. I think there's a good chance this country is developing a good framework for wildlife conservation."

Curio shops are still doing a booming business in zebra hides, elephant tusks and other animal products despite the Kenya Government's total ban on sport hunting.

"Animals have a right to survive as part of the economic base of Kenya," Mr. Kamora said. "I take the optimistic view that wildlife will be there even in a hundred years. We will always have wildlife."

The continuing existence of official curruption, however, still worries many and prompts strong condemnations of Kenya. To be fair, one must compare Kenyan corruption with similar problems in highly profitable smuggling and other illegal endeavors elsewhere.

Look, for example, at the inability of the United States Government to stop heroin smuggling, or even to prevent entire planeloads of marijuana from entering the country. Look at the ease and openess with which prostitution, extortion and narcotics rings operate within major American cities. Few governments could claim to be more efficient, or to have more resources, or to be able to call upon more public support than those in the United States. Even in matters of wildlife protec-

tion within the United States, it took many years for Federal officials to bring egret and alligatior poaching under control. And, in the days when the United States was a developing country, the rate of wildlife destruction was vastly greater than that in Kenya today. The central and high plains of North America once were as abundant with game as East Africa, but Americans decided to grow wheat and corn instead.

How, then, can one expect more from a poor country like Kenya, which has so many other priorities of equal importance such as education, health care and food production? Furthermore, the demand for wildlife products comes overwhelmingly from rich people outside Kenya, from people over whom Kenya has no jurisdiction.

According to Mr. Kamora, Kenya already spends $10 million a year on conservation, a comparatively large share of its budget. "If we were not for conservation," he asked, "why would we be employing 3,000 men to look after animals?"

How does one balance an American's desire to know that there are elephants or leopards or giraffes living free in the world, against an African's desire for enough land to feed his family?

Amboseli National Park is visited regularly by tourists who come to see its wildlife.
Mount Kilimanjaro can be seen in the distance.

And Kenya is preparing to do more. As part of a major new effort, Kenya has pledged an additional $19.4 million through 1981 to be spent in conjunction with a $17 million World Bank loan to develop a stronger antipoaching program and to improve its wildlife-management practices. The money will add three new antipoaching units, and create a "wildlife planning unit" that will establish scientifically sound data on wildlife populations, carrying capacities (the number of animals the land can support) and other facts necessary to a wildlife management policy that can assure viable wild animal populations in perpetuity.

Another cause for hope is the growing popular support for wildlife conservation among Kenyans themselves. In the past many Kenyans, like early American pioneers, looked upon wild animals as a menace, something to be gotten rid of to make the land safe. other Kenyans—the country has dozens of distinct ethnic groups with widely differing traditions—coexisted for thousands of years with wildlife, much as did the Indians of North America.

Now, however, there are millions of Kenyans in cities and densely populated farming regions who have never seen big game animals. Through public education efforts, including the enormously popular Wildlife Club of Kenya, which has chapters in 446 schools, many Kenyans, are adopting proconservation views quite like those of Westerners. "I believe the day will come in this country," said Mr. Jenkins, the Meru warden, "when an appreciation of wildlife for its own sake will be common. There's a big awareness of wildlife, especially among the youngsters."

Despite the impact of population expansion and of poaching, Kenya has not seen the end of its game animals. The game that is ending, however, is the carefree romp that white settlers and big-game hunters had, the get-rich-quick profiteering by newly indepentdent African leaders, and the selfish attitude of Westerners that Kenya and other poor countries were obligated to preserve all their wilderness for the esthetic pleasures of the rich countries.

The end of the game is the end of innocence. It is the beginning of a more mature, more responsible attitude towards all the species of Africa, including *Homo sapiens*.

When tourists visit Kenya's parks, they stay at hotels such as Kiraguni Lodge in Amboseli National Park.

Mr. Yunda Kamora, the wildlife ministry's permanent secretary believes that "animals have a right to survive as part of the economic base of Kenya."

The World's Deserts are Growing

by Boyce Rensberger

The land is dying. Over vast reaches of every continent the rainfall, soil fertility and vegetation necessary to most forms of life are diminishing or disappearing. The result is that the world's deserts are spreading and new deserts are appearing and growing. The rate at which this phenomenon, called desertification, takes place has been increasing in recent years, posing what observers of global ecological trends see as one of the major challenges to mankind's ability to cope with an expanding need for food and space.

It is estimated that fertile, productive land is being denuded and destroyed, in large measure because of man's mismanagement, at a rate of 14 million acres a year. Already about 43 percent of the planet's land surface is desert or semi-desert. Unless desertification can be slowed, some scientists say, fully one-third of today's arable land will be lost during the next 25 years, while the world's need for food will nearly double.

To address the problem on a global scale, a United Nations conference on desertification convened in Nairobi, Kenya in August. Some 1,500 delegates from more than 100 countries and 150 governmental and private organizations discussed the problem, reviewed technical reports prepared in advance and formulated a "plan of action" for marshaling the world's resources to halt or reverse desertification.

CONFERENCE IN NAIROBI

The conference was scheduled in 1974, when it appeared that the long drought in the Sahel, the southern fringe of the Sahara in Africa, might continue indefinitely to push the desert and growing famine southward into more populous regions. The drought ended but left a new appreciation for the impact of desertification in the Sahel and hundreds of other places where deserts are growing or forming.

Long tinged with an air of mystery and romance in Western eyes, the world's great deserts—the Sahara, the Gobi, the Kalahari, the Arabian, the Sonoran, the Patagonian and others—have seemed to be bleak and unchanging environments, otherworldly places with little connection to the verdant lands where most people live.

In fact, recent studies of global climate have shown the deserts to be integral parts of the weather systems that give some regions abundant rain precisely because other regions get little or none. And intensive surveys of the earth's agricultural potential have revealed that while the productivity of arid and semi-arid lands is low, such land is essential to supporting the human race. About 14 percent of the world's people, some 628 million, live in dry lands, almost totally dependent on a marginally

productive environment that is rapidly withering.

Desertification is not a new phenomenon. Many of the places where civilization—that is, agriculture and the beginnings of urbanization—first emerged were destroyed by it. The names of Ur and Babylon, Dilmun and Ebla came from what are now the parched sands of the Middle East, a region once known as the fertile crescent. Like Ozymandias, the ruins of these ancient civilizations are mute testimony to the dependence of human society on land that is biologically productive, and to the possibility of killing that productivity.

Although the Sahel drought, and a worse though less publicized one in Ethiopia, were once thought to be the result of a global climatic shift, the return of normal rains for several years has weakened this theory. Closer study of what happened in the Sahel and elsewhere has demonstrated that a far more potent cause of desertification is the hand of man. Evidence indicates that it was mismanagement of the land, not a climatic change, that doomed the ancient civilizations of the Middle East. Technical reports prepared for the Nairobi conference indicate that many of the same forms of mismanagement continue to be the major threat to human survival in arid and semi-arid regions of the world.

Among the chief causes of desertification are overgrazing by livestock, overcutting of forests, improper tillage for crops and overconcentration of human and livestock activities around scarce water sources or settlements. Even irrigation, if it waterlogs poorly drained soils or deposits accumulations of toxic salts, can kill the land. Such factors have operated for centuries but only in recent decades has the growth in human and livestock numbers intensified the pressures beyond the land's ability to recover. In the past the peoples of the arid lands coped with the limits of their environment through a variety of traditional practices that minimized the impact on the land.

SHIFTING THE BURDEN

Nomadism, for example, moved people and livestock from an area before its vegetation was totally destroyed. By moving almost continually, nomads allowed an area's vegetation to recover before they visited it again. Iraqi pastoralists, to cite another example, used to irrigate a region only every other year, alternating their herds from one pasture to another. This minimized the effect of waterlogging and salinization. Religious practices helped to sustain environments in the Rajasthan region of Pakistan and India. Because trees were held sacred, those that would not otherwise survive were maintained because people watered them regularly as acts of devotion.

Such traditional ways of coping with aridity are rapidly disappearing, largely through the impact of Western technology and ideas. The major factor is, of course, human population growth, partly a result of public health measures that have cut the death rates. Similarly, improvements in the delivery of veterinary services have improved the survival rates of cattle, the chief livestock species in semi-arid zones.

As human and animal numbers grow, pressure on the fragile arid lands increases. When a drought comes, the land's carrying capacity is suddenly exceeded. Vegetation is stripped beyond regeneration and populations crowd into the remaining islands of green, often pushing them beyond capacity, too.

Ironically, one of the major efforts to assist drylands inhabitants during a drought—the drilling of wells—has caused further desertification: So many feet and hooves trample the area near the well that vegetation is destroyed. Combined trampling and grazing has produced steadily enlarging circular deserts around many water holes in arid lands. The need for firewood for warmth and cooking has also caused huge tracts to be denuded of trees that were often the last traces of green and shade.

Documents prepared for the desertification conference note that although the human suffering caused by desertification is greatest in developing countries, destruction of the land is rampant in many industrialized societies as well.

In the United States, for example, overcutting of trees in many areas has stripped the land, making it vulnerable to soil erosion. Strong pressures to increase agricultural production have led ranchers to graze cattle on relatively dry range lands and to plow up fragile prairie grasslands. Both activities destroy native vegetation capable of surviving in arid conditions and leave the soil lifeless and unable to resist

Vast areas of Australia are desert. The Simpson Desert is in the central part of the continent.

Erosion by wind and water has created spectacular desertscapes. Here in Monument Valley, Arizona, buttes rise dramatically from the desert floor.

*The Kalahari Desert in Southern Africa: Evidence shows that
mismanagement of the land has led to severe desertification.*

desiccation and erosion in periods of drought.
In the irrigated fields of southern California,
another form of desertification is under way as
toxic residues from irrigation water accumulate
in the soil.

Among the items high on the agenda of
conferees in Nairobi were a variety of proposals
for preventing or reversing desertification. They
range from the traditional calls for more
financial aid from the world's wealthier countries to specific programs for replanting deserted
areas, lowering the birth rates and livestock

numbers among nomads, harvesting scarce
rainwater and concentrating it and, in some
places, abandoning arid lands entirely.

Like other United Nations meetings on global
issues affecting human welfare, the Nairobi
desertification conference is unlikely to reach
any grand solutions. It does, however, appear
already to be marshaling appropriate bodies of
technical and political attention to a problem
that has never before been addressed as a global
phenomenon that affects most of the world's
countries.

Battle Over the Mighty Colorado

by Grace Lichtenstein

All winter long, snow falls on its mountain-crested rim. . . . When the summer sun comes, this snow melts and tumbles down the mountainsides in millions of cascades. A million brooks unite to form a thousand torrent creeks; a thousand torrent creeks unite to form half a hundred rivers beset with cataracts; half a hundred roaring rivers unite to form the Colorado, which rolls, a mad, turbid stream, into the Gulf of California.

That's how Major John Wesley Powell, who in 1869 was the first white man to explore the unknown Colorado River, described it only a century ago. Today, he would hardly recognize it. Last winter, a fraction of the normal snow fell on the Rocky Mountains, and by June many of the "million cascade brooks" were already dry. But even had the usual snow fallen, the mighty Colorado has not been a "mad turbid stream" for many years.

Instead, this 1,450-mile long river, which carved the Grand Canyon, has been turned by man into the most controlled, most litigated and one of the most important waterways in the United States. Some 17 million people in seven states and part of Mexico depend on its comparatively meager flow for their household needs, their electric power, their food and even

some of the clothes on their backs. It forms the cardiovascular system—the lifeblood—for a huge chunk of land that is both the most arid and the fastest-growing region in the nation.

Right now, in the middle of the worst Western drought in 71 years and the big debate between President Carter and Congress over Federal dam-building, the Colorado's cool streams are provoking great waves of national and international controversy. A journey down the river reveals a technological revolution, through which every single drop of water is used and reused to make deserts bloom. Cities, farmers, oil-shale companies, coal miners and electric-power producers are fighting one another for rights to wring out even more, while environmentalists are battling to stop any further exploitation altogether. And some experts say the river has become a perfect example of how this country wastes its water as recklessly as it has wasted its fossil fuels.

By following its course, in stages by foot, car, plane or boat, from the snowy peaks of Colorado to the alkali salt flats of Mexico, one is struck by the realization of how much of the future of the Western United States depends on who wins these fights. The journey shows, too, why the importance of water is so misun-

The spectacular Colorado River.

derstood in the lush East and is the cause of such frustration in the arid West. As Millie Elkins, a Federal official in Yuma, Ariz., puts it: "Every cupful is somebody's life." And Roland Fischer of the Colorado River Water Users Association fumes over what he contends is the ignorance of Federal officials in Washington: "Those people don't know a head gate from a grease rack."

THE HEAD WATERS

It begins in the high country.

The headwaters of the Colorado are the streams from melting snow in the forests and tundra of Rocky Mountain National Park, whose peaks form the spine of the Continental Divide. It is late spring on a hiking trail in Phantom Valley near a rock formation called Hell's Hip Pocket. I follow a sign at 9,000 feet above sea level through towering blue spruce until I come upon a babbling brook, ankle deep. I sprawl flat on the ground to drink the icy liquid. It tastes better than Dom Perignon. This is the North Fork of the Colorado. It remains in its virgin condition, "wild" and unchecked, for a total of just 12 miles. But even at these headwaters, there are unmistakable signs of trouble—the kind of trouble greatly feared by Westerners. "Normally this water would be boiling down through here," says Bob Barnes, a Bureau of Reclamation hydrographer, but the drought has already reduced it to levels more often seen in midsummer.

So what, who cares? Many people, not least among them David Dodge, manager of the C Lazy U Ranch, East Division. He's the biggest user of water from the Red Top Ditch, which was cut in 1910 as the first man-made diversion from the North Fork of the Colorado. Sixteen miles long, it carries the water to four ranches, which have grooved out smaller ditches that run through their hayfields. The water is free, in accordance with state law that says the right to its use belongs to the first person who uses it beneficially. The Red Top Ditch ranchers have "senior" water rights, which means they've inherited or bought the rights from the earliest users.

Dodge, with his senior rights, is not one of those angling for more water. But this year he

knows that such rights have little meaning. By early July, the ditch is almost dry. He and the others use the water to grow hay that's sold as cattle feed. As soon as the water stops, the hay must be cut prematurely, which may mean a very small yield. What can he do? "Pray for rain," Dodge says over lunch at the ranch house. "Otherwise, the boss will have to go to the bank and borrow money." Multiply the problems of this one ditch at the headwaters of the Colorado a thousandfold, and one has an idea of the plight of ranchers and farmers throughout the drought-stricken West. Not enough hay means reduced income, but it also means a smaller supply of cheap feed for beef. Thus, ranchers have to choose between much more expensive cattle feed or simply selling the cattle for quick slaughter. That, in turn, leads to a glut on the beef market, which may temporarily cut the prices a housewife in Queens will pay for ground round steak. Eventually, however, it will probably mean much less beef on the market, and thus soaring prices.

RESERVOIR BEGINS

The pastures and hayfields of the C Lazy U are just around the bend from Grand Lake, the first major natural lake on the Colorado. It is also the first in a network of natural and artificial reservoirs that the Federal Government and others have utilized throughout the Rockies to capture and store the priceless snow runoff. From these reservoirs, three massive engineering projects pump and push that stored water through intricate siphons and tunnels that carry 27 percent of the state's share of Colorado River water through the mountains to the east side of the Continental Divide. There, it irrigates thousands of acres of once semiarid farmland and provides water to such ballooning cities as Denver, Fort Collins and Colorado Springs.

About 99 percent of the state's population now relies on the river's water, and these diversions across the Divide have created a fierce tug-of-war. Agriculturalists and energy interests in western Colorado are pitted against agriculturalists and big-city water boards in eastern Colorado. The western-slope residents fear that the city folk will eventually "steal" all the water the state is allowed to take from the

58   *The Colorado River begins in Rocky Mountain National Park from streams of melting snow.*

river. "Allowed" is a key word. One of the numerous paradoxes about the Colorado River basin is that although most of its water (in the form of snow) starts in Colorado, Wyoming, Utah and New Mexico, those states are required to let the majority of the river flow into, and be used by, Arizona, Nevada, California and Mexico. A "compact" written in 1922 divided up the scarce water among the states, and they've been squabbling about it in Federal courts ever since.

THE COLORADO BASIN

There would be no need to squabble, of course, if the river were an engorged giant, like the Columbia River in Washington, whose estimated average annual flow is 180 million acre-feet. (What is an acre-foot? The volume required to cover an acre of land with one foot of water, or 326,000 gallons, enough to supply the domestic needs of a family of five for a year or fill 19 average-sized backyard swimming pools or flush 68,631 toilets.) Nor would there be squabbling if there were sufficient rainfall. The trouble is that the Colorado's annual flow is puny—an estimated 15 million acre-feet. This is about the same as the Delaware, which is less than one-third the Colorado's length, drains a much smaller basin area and exists in the wet East. Overall, more than 90 percent of the Colorado and its tributaries is used to irrigate agricultural fields that can expect no help from the clouds. The most important fact to keep in mind about the Colorado River basin is that it drains one-twelfth of the land mass of the United States, and that land mass is mostly desert.

The rapid agricultural growth of the West, the population growth of desert cities like Phoenix and Los Angeles, the growth of energy production all combine to strain the Colorado even more than when the compact was found necessary back in 1922. Proposed coal-slurry pipelines, coal gasification plants and oil-shale schemes are all predicated on the use of that same water. The Colorado water is already recycled for hydroelectric power, irrigation and drainage back into the system about three times as it travels from the headwaters to the Gulf of California. The more the competition for water,

the greater the need for new ways of collecting and storing it. This is a principal reason why Colorado has been fighting President Carter over several dam projects he wants cut from the Federal budget.

Colorado as a whole is concerned that states farther down the river—especially Arizona—will soon "steal" its water. The Central Arizona Project, now under construction, which would take a mammoth drink of the river when completed in 1985, was approved years ago only after Colorado made sure the legislation included extra dams within *its* boundaries. But two of those proposed dams seem likely to be killed by Congress in accordance with Mr. Carter's wishes.

Philip L. Fradkin, author of a forthcoming book about the river, believes the water-projects battle is really part of what he calls the "second major land-use revolution" in the West. During the first revolution, open range gave way to fenced-in, high-crop-yield farming in the late 19th century. "The second major change now underway is an irrevocable change from a rural, agricultural society to an urbanized, energy-producing one," he said recently, adding, "This change would be impossible without the waters from the Colorado River. The river basin is symbolic of the whole West. It's the last frontier in the lower 48 states."

GLEN CANYON

From Grand Lake high in the mountains I follow the river by plane down to the Glen Canyon Dam area on the Arizona-Utah border. What I see is just how low the Colorado is, and how engineers have tinkered with the stored water as if reservoirs were cups with spigots on their sides. The engineers have opened and closed different spigots, spilling or pumping water from one "cup" to another, some reservoirs being deliberately drawn down to fill others. Thanks to the large amounts stored in the reservoirs from Wyoming to Arizona, there's enough water to fill most needs this year. But if the drought continues for a second and third year, if the snowfall next winter is again scant, all these cups will be nearly drained. The entire Colorado system has 14 major dams backed by reservoirs that can hold a three-year average flow.

The Colorado's annual flow is low while its basin area is enormous.

Grand Lake is full when I fly over it, but Lake Granby, a little farther down-stream, is half empty, its barren shoreline now strewn with rocks and marooned pleasure boats. Nevertheless, the river's plumbing system seems in working order, its flow regulated not by nature, as in John Wesley Powell's day, but by computers in the small town of Montrose, Colo. The computers are programmed to determine when the various head gates and power generating turbines in the upper river dams should release water and when to hold it back.

Trace the river as it collects tributaries from Wyoming, Colorado, New Mexico and Utah, and some remarkable aspects of the Colorado become easier to understand. The Colorado has

never been used for transportation the way the Mississippi or the Hudson has. It was too wild— "too thin to plow, too thick to drink," as boatmen put it. And it ran through such sparsely populated, rugged landscape that no major cities or ports developed on it. There is still almost no industry along its banks, and thus little industrial pollution. In fact, most of the Colorado is no longer red, as its Spanish name implies it should be. The huge buildup of silt that once colored it has largely been filtered out. The major pollutants in the river and its major tributaries, like the Green River that joins up in southeast Utah, come from motorboats on the reservoirs and from cattle. "Ranchers have been bulldozing manure into the Green," explains a

Bureau of Reclamation spokesman. "Our environmentalists are shocked. They're afraid Lake Powell [the reservoir above Glen Canyon Dam] will be ruined in a few years.

One of the few sections that now bear any resemblance to Major Powell's eloquent descriptions is a short stretch of foaming water called Cataract Canyon—through which I float by raft from plunge to plunge of roiling rapids as the river cuts through the splendid red rock walls of southeast Utah. All too soon, these rapids give way to the serene blue waters of that controversial lake named in Powell's honor. The good major probably would have been horrified. Despite the strenuous objections of conservationists, Glen Canyon Dam and Lake Powell were created in the 1960's by flooding a beautiful natural canyon.

The Glen Canyon Dam is an imposing curved shield of concrete that looks out of place amid the grotesquely exquisite sandstone buttes, knobs, mesas, canyons and natural bridges of southern Utah and northern Arizona. A few miles downstream, in a lovely little delta greened by tamarisk, is Lee's Ferry, the historic site where an exiled Mormon murderer named John D. Lee first began transporting passengers in 1871 across one of the few spots on the river not hemmed in by forbidding high cliffs. Since 1922, Lee's Ferry has also been the spot at which the annual flow of the river is measured, the dividing line between the "upper" and "lower" basin sites.

The dam has eight power plants inside it that can generate 900,000 kilowatts of clean hydroelectric power. The Government ordinarily sells that power to utilities all over the West. The income, over the years, pays back the cost of constructing the dam. Unfortunately, the drought has thrown a monkey wrench into this neat system, to evoke a favorite phrase among environmentalists. (Edward Abbey's novel "The Monkey Wrench Gang" is about a band of ecological saboteurs who plot to blow up Glen Canyon Dam.) So far, in 1977, the Federal Government has had to tinker so much with its Colorado River dams that it has disrupted wildlife and river recreational activities. To fulfill its contracts, the Bureau of Reclamation had to buy expensive extra power from oil-fired generating plants elsewhere. The bill came to a whopping $28 million, and the bureau had to ask Congress for a special grant to pay it.

THE GRAND CANYON

Just below the Glen Canyon Dam lies another of the few sections along the Colorado River that has retained its earlier character. Grand Canyon still belongs to the esthete and the naturalist, to the red-tailed hawk and the mountain sheep. The river itself is not majestic like the Hudson, nor thrilling like the Salmon. It is simply a body of water that takes rafting parties through the most beautiful scenery on the continent—that is, when there's water. As it happens this year, the Glen Canyon Dam was kept almost closed from April through mid-June until the river below was reduced to a trickle. Despite warnings to that effect, hundreds of vacationers were stranded in the canyon as the river literally dried up beneath their boats. Helicopters had to fly food in to some, and rescue others. The dam officials finally had to release some water to "flush" some boats out.

Normal amounts of water have been released since then. But others are worried about being left high and dry. *Upper* Colorado River utilities and ranchers have threatened to sue the Bureau of Reclamation if it lets *too* much water out of Lake Powell. They say the lake might then have to be refilled with upstream water that rightfully belongs to them. Five years from now, the upper users argued, they might be denied the chance to take out normal amounts of water for irrigating their hayfields. All this occurred because not enough snow fell during the winter on those peaks in Rocky Mountain National Park. "We're not going to sue God," commented Rollie Fischer, speaking for upper Colorado ranchers, "but we sure are unhappy."

HOOVER DAM

As I tour the Hoover Dam and the stretch of river below it, another irony of the Colorado is apparent. The drought, it seems, has been a bit of a blessing in disguise for those living below that gigantic cork plugging the river southwest of the Grand Canyon. From Hoover south toward Mexico, numerous people, though aware of the hazards, have built second homes, restaurants and tourist facilities in what was once the flood plain for the river along the California-Arizona border. Then, in December

Though the Colorado's water is dammed, used and reused, it still brings rafters through some of the most gorgeous scenery on earth, as in Arizona's Marble Canyon.

The tame Colorado is plugged by the Hoover Dam.

The wild river snakes through a wintry southern Utah.

1976, the Bureau of Reclamation warned them that it probably would have to start releasing much more water for the next 10 years, perhaps flooding those people out. The reason? Lake Mead behind Hoover Dam was at capacity; Lake Powell was almost full after its 13-year buildup, and the Central Arizona Project was not ready to draw out its hugh allotments. It looked as if there would be too much water in storage.

The drought arrived just in time. So Manuel Lopez Jr., regional director of the lower Colorado for the Bureau of Reclamation, has a new worry. Water from the Arizona project is to be drawn out of Lake Havasu, 163 miles downstream from Hoover, in 1985. The amount: 1.2 million acre-feet each year. If all the reservoirs above are not full, Lopez acknowledges, "it might be difficult to meet all our water commitments in the lower basin." Who holds those commitments? Not surprisingly, the biggest share belongs by law to the state that contributes the least to the Colorado River — California.

CALIFORNIA'S STAKE

To understand the relationship between California and the Colorado River, imagine a University of California student spending a January ski holiday in Aspen, Colo. Then, during spring recess, she drives with friends to the Parker "Strip," the Sunbelt's Fort Lauderdale on the California-Arizona border, for some partying. When she returns to Los Angeles, she turns on her air-conditioner, takes some fresh asparagus from the refrigerator and rinses it in preparation for her dinner. The snow she skis on is, basically, the same water she swims in along the Parker Strip, the same water that powers her air-conditioner, the same water that grows the asparagus and the same water in which she washes it. Nearly all of it is water from the Colorado River.

Although recreation was not originally recognized as a "benefit" of the building of Hoover Dam in the 1930's, it and other big dams on the river have created watery playgrounds for some nine million vacationers a year. They spend an estimated $45 million annually, making boating and swimming almost as much a moneymaking venture as hydroelectric power, which produces some $80 million annually for the Federal Government in gross revenues. Two-thirds of Hoover Dam's hydro power goes to Southern California. Half of *that* power is used to run the pumping plants downstream on the river to pull the water up into siphons, over mountains and through 300 miles of desert along an aqueduct that takes it to the lawns and faucets and air-conditioners of Los Angeles. This year, the dams have been crucial to the survival of the California life style that Colorado River water has made possible. "This system has bailed California out," says Mr. Lopez.

Earlier this year, when the second-year effects of the drought in California began wringing northern California dry, state and Federal engineers worked out a plan. In a complicated shuffle, Los Angeles gave up some of its northern California water temporarily, and arranged to get more water from the Metropolitan Water District, a "wholesaler" that sells water to southern California cities. The M.W.D. in turn gave up some of *its* northern California water, too, in return for the right to take nearly double its usual share from the Colorado. The net result means that hard-pressed northern California can keep what little is left of water in its vicinity. Meanwhile, Los Angeles and San Diego are getting saltier, harder water from the Colorado. On Lake Havasu, the M.W.D. has three thick siphons that suck up the Colorado water for its trip westward. Now, for the first time, the M.W.D. has all of its nine pumps attached to the siphons working at full speed.

"The big difference is water quality," said a spokesman for the utility. "Colorado water is saltier and people have been used to getting a blend of it and purer northern California water. The water's harder now and you have to use a better detergent." He also acknowledged that the extra pumping is a financial windfall for the Metropolitan Water District. It pays a pittance to the Federal Government to pump the water in the first place, and now it's getting extra revenue from the increased supplies to those thirsty coastal cities.

The bargain-basement rates California pays for Hoover Dam power is another sore point among neighboring states. The Metropolitan Water District last year paid $3.9 million for $1.3 billion kilowatt-hours from Hoover. The city of Los Angeles pays correspondingly cheap

rates. By contrast, buyers of power from Glen Canyon Dam on the *upper* Colorado pay nearly twice as much per kilowatt-hour. Why? Timing, political pull and inflation. The California utilities signed 50-year contracts for their power in 1937, in the midst of the Depression. The dam itself was much less expensive to build than Glen Canyon in the 1960's.

A decade from now, those California contracts expire, and neighboring states are rubbing their hands in glee, expecting the Government to raise its Hoover prices accordingly. "We expect all hell to break loose," an M.W.D. spokesman admitted. But experts believe California might still get a good deal on new power contracts because it has always had clout when involved in negotiations over the Colorado River. Under various decrees, California emerged with the biggest amount of water allowed any state each year—4.4-million acre-feet. It won a court battle with Arizona when the latter state tried to halt the building of Hoover Dam, knowing it would benefit California. (In 1934, the Governor of Arizona called out the National Guard to stop Californians from building Parker Dam, holding up construction for 11 months.) While the other states were still infants, California had private utilities that could raise the capital to build the waterworks that bring the Colorado River water hundreds of miles to Los Angeles, San Diego and the Imperial Valley. In the current Congress, three California representatives—Harold T. Johnson, John J. McFall and B. F. Sisk—dominate legislation governing water projects. One thing is fairly certain. If the M.W.D. rates go up 10 years from now, the price southern California residents pay for their water and electricity are bound to spiral, too.

LAKE HAVASU CITY

Kilowatts and acre-feet are what the dams and siphons along the lower Colorado are all about. But they're far from the minds of the older folks at Lake Havasu City, just north of the siphons, or the California teen-agers camping on the Parker Strip farther south. Once Hoover Dam was built, engineers continued damming and channelizing the formerly raging river. There are now seven more dams below

Hoover, smoothing out the torrents and leaving a placid blue stream that bulges into reservoirs in spots. These changes made the river banks ripe for development. Fourteen years ago, the McCulloch Oil Company began staking out lots for a new town, Lake Havasu City, on a bleak outcropping of unvegetated desert along the Arizona side. Meanwhile, more and more boating enthusiasts each year discovered that the river channel just below Parker Dam farther downstream was ideal for splashy motorboat races. California youths also discovered that the drinking age on the Arizona side of the river was 19, not 21 as in their home state. Nowadays, Lake Havasu City and the Parker Strip stand out as symbols of the tamed river's appeal. They are also further proof that in the arid West, water is the key element in profitable economic ventures.

A Phoenix librarian visiting Lake Havasu City for the first time not long ago took one look and described it as "a great big Kitty Litter divided by blacktops." It is a city waiting to happen. Everything is there, except most of the people and most of the houses. The town has nearly 29,000 single-family home lots planted in the desert. With some high-pressure salesmanship, the McCulloch organization sold most of them at prices ranging from $6,000 to $34,000 a quarter-acre. The initial attraction was the city's access to boating and fishing on Lake Havasu, created by the Parker Dam. Today, fewer than 4,000 actual homes have been built. Fresh blacktop roads snake through the desert, with nothing on either side but a yellow fireplug here and there. The town does have a population of 14,000, but not many people are in evidence in the summer when the thermometer hovers around 120 degrees. The "city" is either a valuable real-estate development taking shape more slowly than anticipated, or a possible land-fraud scheme, depending on whether you talk to the developer or to the county attorney. (He's now investigating Havasu sales practices.) It's hard to figure out which is more incongruous: the town itself, or its most famous "landmark," the London Bridge, imported from Britain by McCulloch and rebuilt across a finger of the lake as a tourist attraction.

Lake Havasu City, with many retired persons among its residents, is languorous. The Parker Strip, with its hordes of youthful Californians crowding the Arizona-side campsites and tav-

erns every weekend, is "a zoo," in the words of Ross Carpenter, an Arizona park ranger, who says that the youths come there mainly to get drunk and have sex. An estimate 1.5 million of them show up each summer weekend. Even more are there the first weekend in March, when the Parker Enduro, "the Indy 500 of boat racing," takes place. One regular, Rick Patynik, a landscape contractor from Whittier, Calif., tried to explain why he towed his $11,000 white Lavey Craft motorboat on the five-hour drive to Parker so often. "It's the great American dream," he said, sitting on the grassy beach, shouting over the noise of the motorboats zipping past. "It's getting away from L.A., getting away from the rat race."

THE IMPERIAL VALLEY

Floods are hard to imagine on the now tranquil lower Colorado. Yet, Imperial Dam, still farther south on the California-Arizona border, owes its existence to the floods that regularly inundated the desert around it at the start of the century. Imperial was built after Hoover to divert as much of the Colorado water as feasible into a canal that led to California's waiting Imperial Valley. Today, that valley is Exhibit A in any discussion of irrigated agriculture. Once a worthless tract just north of the Mexican border, it is now a booming swath of farms producing two and a half bumper crops a year of everything from asparagus to cotton. Imperial is the nation's fourth largest agricultural country. Every asparagus stalk, every melon, every cotton boll in the vast district grows there thanks to Colorado River water. The valley's irrigation district is the single biggest beneficiary of the river—2.9 million acre-feet a year. "We don't want rain here; it does more damage than good," remarked Robert F. Carter, general manager of the irrigation district, as he drove along the sunbaked fields, each one bisected by arrow-straight canals of water, flowing in from the river 160 miles east. "The river's our sole source," he continued. "Take our water away, and you'd be taking $525 million worth of commodities from the tables of the people of the United States."

Since 1900, millions of additional acre-feet of water have been diverted from the river to irrigate hundreds of square miles more of previously uncultivated desert in California, southern Arizona and Mexico. Indeed, the Mexican-built Morelos Dam diverts the last remains of "real" river water to irrigation canals. What is left of the "Colorado" after Morelos is actually drainage water that has seeped through the soil and returned to the river bed after doing its job among the crops. At this point, down near what river people call "the end of the ditch," every irrigation district, state agency and Federal official watches the diversions very, very carefully. Every twist on the spigot of a canal head gate means dollars to farms, business for long-haul truckers, tax revenues for states, hay for cattle herds, food for dinner tables from Los Angeles to New York and cotton (California is the biggest cotton-growing state, Arizona the fourth biggest) for Americans' clothes.

In the Imperial Valley, 140 *zanjeros*—"ditch-riders" employed by the irrigation district—patrol the thousands of miles of subsidiary canals that serve each farm. They make sure that each farmer gets his fair share. It's an immense job; if all the Imperial water ditches were laid end to end, they'd stretch from San Diego to Manhattan. In Arizona, ditches in the ever-expanding, irrigated farmland fanning out from Yuma are similarly patrolled.

IT FLOWS TO MEXICO

Mexico watches the water as well. The United States is obligated to let 1.5 million acre-feet of the Colorado River flow across the border. In the 1960's, so much additional water was diverted in Arizona to newly irrigated farmland, it made the water that Mexico took at Morelos Dam saltier and saltier, because of the additional soil it passed through. The saline water ruined valuable crops in the Mexican fields. Each year, Mexican authorities complained louder and louder. Finally, in 1972, President Nixon called on Herbert Brownell, former United States Attorney General, to solve the dispute. Brownell's answer: build a desalting plant at Yuma, where the United States could clean up the water before it went into Mexico. The cost of the desalinization plant and related facilities, with the first contracts to let this

summer, is now estimated at $316 million. In other words, Colorado River farmers overused the river water so much that the rest of the country's taxpayers must pay to correct that abuse.

One mile above Morelos Dam, in a rather deserted sandy spot along the Colorado lined with willows, the ultimate river surveillance takes place almost every day. Alton Clark, a United States civil servant, rows out to the middle of the murky greenish stream in a red rowboat, dips a bottle in it and collects a sample of water. He's measuring its salinity. At the same time, Elmer Carlson, another U.S. civil servant, pulls himself out over the river in a handcart hanging from a cable 40 feet above the river and dunks a device that looks like a lead fish hanging from a thin steel-cable fishing rod into the river at 20 different spots. Each time, he puts a pair of earphones on while he peers intently at a stopwatch. He's measuring the precise amount of water flowing that morning into Mexico. It takes about an hour. The United States Boundary Commission in Yuma employs three men to perform this task. Each makes about $16,000 a year. The Mexican Government employs three men to perform the same duties on alternating days. "We don't send any more water down to Mexico than we have to," explained Henry C. (Spike) Ganster, the supervising engineer. And that's just how precious the Colorado River water is to two sovereign nations and their farmers.

INDIAN WATER RIGHTS

There are other sovereign nations keenly interested in the Colorado River as well—five central Arizona Indian tribes. They're afraid that they won't get *their* fair share when the Central Arizona Project is complete, carrying water from the river to interior farms and the cities of Phoenix and Tucson. To avoid the Federal courts, where Indian water-rights cases clog all court calendars, the five tribes got Senator Edward M. Kennedy to introduce a bill specifically granting them a substantial amount of the river's flow. A controversial section of the bill specifies that water would go to the Indians by taking it away from the most recent patch of desert-turned-farmland in southern Arizona—the Wellton-Mohawk district of 151 farms.

Arizonans are, predictably, up in arms over the idea. Take away that water, and the district would revert to desert. An aide to Senator Kennedy explained that the bill was intended to protect Indian resources, as well as to replace endless court battles with a legislative solution to Indian water disputes. The aide also pointed out that taking the district out of cultivation would displace only 1,000 white farmers, while giving water to 30,000 Indian farmers. In addition, removing water from the Wellton-Mohawk district would solve much of the salty water problem, eliminating the need for the multimillion-dollar desalinization plant.

Tommy L. Long, a crusty old farmer who has lived in Yuma all his life, spoke for his neighbors when he said, "We have some of the most productive agriculture in the country. Why degrade it in order to give water to somebody else?" Standing near one of the numerous canals in Yuma, Mr. Long, manager of another irrigation district in the area, ruminating about how the river had changed. "I know this river like the back of my hand," he began. "And I can remember when it was a raging current here. When I was a boy, I saw half of Yuma under water, from a break in the levee. I remember when there was a whirlpool here that was the scariest thing you ever saw." What about the dams? "They were essential. If it weren't for the dams we couldn't survive even a two-year drought." He pointed out that, thanks to the dams, the region, with its year-round growing season, has hardly felt the drought.

THE FUTURE OF THE COLORADO

How many new dams should be built in the arid West, what to do about Indian water rights, whether it's smart to grow water-intensive crops like alfalfa in a desert valley, what will happen to the salinity and flow of the river when the Central Arizona Project goes on line, how much oil shale and coal gasification the West could sustain without taking water away from its farmers . . . all these debates will grow more intense in the coming decade.

Westerners feel that Easterners don't understand the dry region's dependence on water. "A Georgia boondoggle is a vital Colorado project," Gov. Richard D. Lamm of Colorado said angrily once, referring to President Carter's antidam stance. And Roland Fischer of the Colorado River Water Users Association declares that "there are people in Washington, D.C., eating irrigation-grown strawberries who have no idea where they came from." Robert F. Carter of the Imperial Valley district says New Yorkers should realize that "starting in 1978 they might not be able to buy lettuce or asparagus because no more land can be irrigated by the Colorado River. I think it's going to happen, if we don't use water prudently."

Environmental groups such as the Sierra Club and Environmental Defense Fund agree with the last statement, but offer a different solution. No more desert land should be irrigated, while farmers should be forced to stop wasting their water. (Only recently has the Imperial Valley district lined its canals with concrete to reduce seepage. In Arizona, there are no meters on farms that use well water, although the farmers use far more water than metered homeowners.)

Will the not-so-mighty Colorado keep producing enough "liquid gold" to fill the needs of the swelling Western Sunbelt? I wonder, as I taste the warm, somewhat brackish flow on the Mexican border, so different from the delicious snowmelt I had sampled 1,400 miles northeast in Rocky Mountain National Park. I wonder even more as a small plane takes me south into Mexico. Millie Elkins of the Bureau of Reclamation points downward to the narrow brown creek, next to which is a wider, bright blue canal. The creek is the Colorado River, the wide canal is a Mexican irrigation ditch. Further south, the Rio Hardy, one final tributary, joins the narrow creek. The land on either side is filled with green geometric farms, but soon they give way to barren desert. Still, some fishermen stand along the banks, using the river to its very end. About 65 miles south of the border, the brown rivulet begins to merge with the white flats all around. "That's what you call a ghost of a river," says Millie Elkins, as we watch the flow peter out and disappear a few miles short of the Gulf.

The mouth of the Colorado is filled with sand.

A Plea for a Mundane Mollusk

by Janet L. Hopson

There are surely better ways to spend an afternoon than sloshing through Big Darby Creek, in Columbus, Ohio, searching for the northern riffle shell. Another wildlife enthusiast, for example, might choose to while it away looking for Attwater's prairie chicken, the unarmored three-spine stickleback, the pig-footed bandicoot or even the Santa Cruz long-toed salamander. All these creatures share something in common besides their ungainly names: They are among the 250,000 species of animals and plants whose very existence is endangered by man's encroachment upon their habitats, and are thus candidates for protection by the United States Government.

Endangered species, by definition, are hard to find, and I have come several hundred miles to this stream in Central Ohio where the northern riffle shell is making one of its last stands. Somewhere in its shallow gray-green waters—nestled among the glacial cobblestones, bearded with algae, filtering food through hairy gills and generating occasional strands of slime—is my quarry. A clamlike fresh-water mollusk—by proper name, *Epioblasma torulosa rangiana*—the riffle shell is small, squarish and, except to experts like the two zoologists who are my patient guides today, virtually unknown and thoroughly unremarkable.

The Northern Riffle Shell

I am knee-deep in the Big Darby, in fact, precisely because the riffle shell's hopeless obscurity helps illustrate the enormous problem of protecting all the species that scientists consider endangered. Public attention thus far has been focused on creatures such as wolves, bears and eagles that are closest to our anthropomorphic hearts. But perhaps only 10 percent of the world's endangered species have sad eyes, lustrous coats or noble wing spans. The rest, including mundane mollusks like the riffle shell and the small, unimpressive plants with the unpronounceable Latin names, go unnoticed and largely unchampioned. Their very obscurity is an extinction force as pressing as pollution, commercial exploitation or the destruction of their habitat.

"Look for the *Epioblasma* in shallow, fast water," cautions one of my guides, David Stansbery, as he plunges into midstream with a seine. A compact, middle-aged man, Dr. Stansbery is director of the zoology museum at Ohio State University. He has spent nearly 25 years studying ecology and especially the group of fresh-water mollusks known as naiades, of which the riffle shell is a member. Now he and his colleague, Dr. Carol Stein, have a contract from the Federal Government to survey the status of naiades in the Eastern United States. Dr. Stansbery seems delighted at the interest shown by this Washington visitor. "It's been tough sometimes getting my research jollies from a group of organisms most people have never heard of," he says.

Dr. Stansbery has been telling me that, surprisingly, there are more types of naiades in North America than there are kinds of mammals. The riffle shell has been around for some 10,000 years, and once was found all the way from Ontario to West Virginia. But after human civilization took over, the fast shallow streams it favored were channeled, deepened, dammed, contaminated with sewage, turned toxic by pesticide run-off—and the riffle shell retreated. In fact, some 40 percent of the 108 naiad species native to Ohio are now extinct or near extinction.

PROFIT VERSUS PRESERVATION

"You are looking at the last free-flowing river in Central Ohio," Dr. Stansbery says with emphasis. "This is what I call a living museum." But, like less-fortunate streams, the Big Darby and its riffle-shell inhabitants now await man's pleasure. Both the United States Army Corps of Engineers and the City of Columbus want to build back-to-back earthen dams and two reservoirs—the engineers for flood control, the fast-growing city for additional water supply. The dams would flood the last breeding population of riffle shells out of their home.

Dr. Stansbery's colleague, Carol Stein, who is hunched over, picking up tiny river snails and putting them in a coffee can, has fought to preserve the creek from encroachment for nearly a decade. Curator of Gastropods at the zoology museum, she now heads a vocal citizens' group opposed to damming the Big Darby. "By presenting the facts long enough and loud enough," she is saying, "I think we have convinced the city it should use the well field [south of Columbus] first." But she is not optimistic: "Nobody yet has found a way to preserve a river from people who could make a profit." On this sunny afternoon, however, the creek looks wonderfully undisturbed, 60 feet wide where we are, cutting two sinuous channels through muddy gravel bars and stands of water willow. Indeed, the riffle shell's chances for survival appear better here in the Big Darby than in the Big Bureaucracy of Washington that ultimately controls its fate.

To qualify officially as an endangered species—and thus stand a reasonable chance of withstanding civilization's onslaught—the riffle shell must first survive a flood of red tape and paperwork. The protection of endangered plants and animals was mandated under the Endangered Species Act of 1973. This legislation, together with earlier laws and an international treaty, provides the authority to list native and foreign species as endangered, and the enforcement muscle to safeguard a species and its habitat from hunters, collectors, dam-builders and bulldozers. The act's strongest provision, Section 7, aims at protecting endangered species from their greatest threat, the Government itself, insuring that the Corps of Engineers and departments such as Agriculture, Defense and Interior do not carry out or authorize projects that would harm the species. Theoretically at least, this provision makes it possible to change or prevent any land-use project where Federal funds or permits are involved—from dams on Big Darby Creek to highways in Mississippi and coal mines in Montana.

The enormous power and responsibility of implementing the act resides with the Office of Endangered Species in the United States Fish and Wildlife Service. Recognizing this responsibility—and subject to intense pressure from special interests in and out of the Government—the office has taken "a conservative approach," in the words of its program chief, Keith

Schreiner. It has set up a formidable obstacle course that must be cleared before a species receives official sanction of its endangeredness. The procedure consists of at least 10 distinct hurdles, beginning with formal notification of the office that a species appears to be in danger of extinction. Opportunity for careful consideration, as well as the possibility for bureaucratic entanglement, is built into practically every step. Step 6, for example, requires that, after a staff scientist has drafted proposed rules and usually an assessment of the environmental impact of protecting the particular species, his report must carry no fewer than 12 signatures from department heads and managers within the Department of Interior. The entire process takes a minimum of 36 work days, according to Keith Schreiner, and several of the steps provide

Grizzly Bear

for waiting periods of up to 90 days before the next step can begin.

Agency officials defend the cumbersome process on the grounds that legally tight cases must be built in order to stand up in court. Indeed, two important cases already have been tested in the courts, both involving the final and most crucial step in the process of designating an endangered species — legally establishing the habitat critical to its survival. In one case, the judge accepted the office's evidence against a challenge by the United States Department of Transportation and stopped construction on a stretch of highway through the last habitat of the Mississippi sandhill crane. In the other case, the judge permitted the Tennessee Valley Authority to go ahead with its Telleco Dam though it threatened the habitat of the snail darter. And there is almost certain to be a prolonged court battle over the office's definition of the grizzly bear's critical habitat, which involves valuable grazing and timber land and, beneath it, rich deposits of natural oil, gas, coal and other minerals.

One result of the office's caution is that, in its first two years, it certified only 11 animals as endangered. Conservationists charged it with "benign neglect" of the nearly 250,000 other species now believed endangered. Staff members, disgruntled over repeated rejection of their listing proposals on the basis of "insufficient documentation," leaked internal memos to the newspaper columnist Jack Anderson. Anderson wrote a column charging that the office was deliberately stalling on half a dozen species, including the chimpanzee and the American crocodile. Congressional committee hearings followed, and the paper-work suddenly speeded up. In fact, almost 200 species have now been officially listed as endangered since the first hearing was called.

Unhappily for its enthusiasts, the northern riffle shell is not among them. Some four years ago, the riffle shell succeeded in clearing the first hurdle by coming to the office's attention: The staff malacologist, or mollusk expert, Marc Imlay, started a file on it and notified the chief of the office that the riffle shell probably merited an official review. During the next few months, he added to the file a few letters of evidence that were prepared by the on-the-spot experts, Drs. Stansbery and Stein in Ohio. After that, the

riffle shell got lost in the shuffle. When I expressed an interest in the creature in the summer of 1976, the most recent entry in the file was a letter from Dr. Imlay in response to an inquiry from state environmental authorities in Ohio. Yes, indeed, he assured them in 1974, the northern riffle shell was a likely candidate for the endangered list. Dr. Imlay told me he had been "bogged down with paperwork on other species." Certainly, the shortage of staff and funds is one obstacle to the riffle shell. The office has to make do with a staff of eight middle-level managers and seven full-time biologists and an annual budget of $10 million, only 5 percent of the Fish and Wildlife Service's budget. At the present rate, says Keith Schreiner, "it will take us the next 6,000 years just to list the endangered species, not to mention developing recovery programs for them."

SOME SPECIES ARE MORE EQUAL

There is also the problem of priorities. Obscure species like the riffle shell can't compete with exotic creatures such as the grizzly bear, the kangaroo or the condor, which are also in imminent danger of extinction. The latter are not only more appealing, they fit more comfortably in the Fish and Wildlife Service's traditional concern with the protection of fish and game—what one Government conservationist calls the "webbed-foot-forked-horn-adipose-fin syndrome." Long-term staffers of the service aren't all that eager to protect other forms of wildlife, says Lee Talbot of the President's Council on Environmental Quality, "when it comes to posies, 'dickey birds,' bugs and shells." Nor, it might be added, is the average citizen who has never heard of the northern riffle shell, and couldn't care less. So why should Washington bother with it at all?

Here on Big Darby Creek, where we are searching for the elusive riffle shell, my guide, Dr. Stansbery, explains the mollusk's predicament. To begin with, he points out, the riffle shell and other naiades weren't always so apparently useless. The early Indians savored their meaty core and made them a mainstay of daily diet. In the 19th century, European settlers built a button industry around their

California Condor

lustrous shell linings. Then, hog farmers found they made good fodder, and an enterprising Japanese businessman discovered the small beads ground from the shells made ideal centers around which to culture pearls. More recently, several companies in the American South showed a brief flicker of interest in naiades, considering the possibility of embedding the pearly shells in transparent plastic for use in paperweights and trivets. The irony is that, while past commercial exploitation undoubtedly contributed to the riffle shell's decline, its present lack of utility may guarantee extinction. Why should the Federal Government preserve a

stream for an apparently useless creature that we can't even find this afternoon? We have been looking for almost two hours, and though our buckets now contain 15 different kinds of shells, our endangered riffle shell isn't among them.

"How do we know what they're good for," chimes in Dr. Stein. "Who knows what we'll find in the future?" Recent science, she notes, brims with serendipitous discoveries involving such lowly organisms. Some kinds of lawn fungi have been found to cleanse the air of carbon monoxide. One group of microbes, which grow in hard, dusty colonies and smell exactly like dirt, are now used to produce antibiotics such as streptomycin and tetracycline. And some molluskan relatives of the riffle shell produce a chemical substance known as mercenene, which apparently is effective against certain forms of cancer. "Each organism contains unique genetic material," says Dr. Stansbery, "and is an irreplaceable natural resource." He is bent over, searching now in a muskrat midden. The riffle shell is not completely useless. Muskrats and other animals feed upon it. "We do know," he says, "that with the extinction of every obscure species, the food web is affected and our own existence as a species loses another element of security."

It is this sense of the interdependence of all living things that has forced many conservationists to conclude that the present approach of protecting individual species may not be the best one. They favor protection of complete ecosystems, large habitations that would preserve, as Russell W. Peterson, former president of the Council on Environmental Quality, explains it, "an entire system of relations, beginning with bacteria in the ground and extending to the loftiest Douglas fir." Obviously, the loss of one species of mollusk is not likely to have major impact on the ultimate survival of the human race. It would be, as Dr. Stansbery puts it, "mainly a psychological loss." The riffle shell is just one of many species on earth: Some two million species of animals and plants have been named and described by scientists (though there may exist several times that number). Some wildlife experts predict the demise of fully one-third of these species by the year 2000. Even if such a forecast is unduly alarmist, naturalists simply don't know enough yet to predict what effects the loss of even a few thousand species of plants and animals will have on nature's delicate equilibrium. Protecting them is an act of faith, a down payment on a healthy ecosystem for the future.

Of late, things seem to be looking up a bit for the riffle shell and the rest of the silent endangered majority. Since Congress started prodding, nearly 2,000 new species have been formally proposed for official certification as endangered. More than four-fifths of these are little-known plants and invertebrates like the riffle shell—many so obscure that, in starting files on them, Government scientists had to assign them common names for the first time.

Mississippi Sandhill Crane

And the word from Washington is that, by March 1977, the riffle shell's own file may be dusted off and started on its certification journey through the bureaucratic maze. Meanwhile, the riffle shell apparently is still alive and well in Big Darby Creek. A cry of discovery goes up from Dr. Stein as she sifts through a heap of muskrat garbage. "Aha! *Epioblasma torulosa rangiana!*" She is holding two northern riffle shells. Both are cracked and the pearly linings are weathered away, but their squarish outline and olive green coloring is unmistakable. "At least," she says in wonder, "the muskrats can still find them."

73

Prairie Chicken

Australia's Biosphere Reserves

by Walter Sullivan

A new concept of conservation that seeks to facilitate evolution, rather than merely preserve the status quo, is gaining momentum in Australia and elsewhere in the world. It would create "ecological reserves" that, insofar as practicable, preserve such life forms as the kangaroo and the other marsupials of Australia, but also permit the slow process of evolution to continue. The reasoning is that nature has never stood still and never will; that life's vigor has depended on continuing evolution, rather than stagnation. In the belief that human activity has become so pervasive that it precludes natural evolution in exploited parts of the world, it is proposed that suitable reserves be set aside for that purpose.

A start in that direction has been made in a number of countries, including the United States, as part of the worldwide "Man and the Biosphere" program. The latter is sponsored by the United Nations Educational Scientific and Cultural Organization.

As stated by Dr. W. L. D. Ride, former head of Australia's Commonwealth Biological Survey, "Land utilization in Australia is rapidly approaching its inevitable end—the day on which there is no land left uncommitted to man's use. On that day the only virgin bush left will be that which has been specifically set aside to conserve it." In the belief that only a system of carefully designed reserves can ensure the long-term survival of this country's exotic and diverse forms of life, there has recently been an explosive growth of its national parks and reserves.

While they are called "national" parks, under Australia's Federal system they are set aside and run by the states, each according to its own ground rules. The rate of their growth is indicated by the increases in national park acreage. In Western Australia, from 1960 to 1975, it grew from 325,000 to 5,675,000. During the same period in New South Wales, it increased from 2 million to 3.5 million. In South Australia the total, including "conservation parks," jumped from 3 million to 9 million. Somewhat more than 2 percent of the country as a whole is now set aside in this manner, although ecologists regard a number of the parks and reserves as being primarily recreational and far from adequate for species preservation or evolution.

A STANDARD FOR MEASUREMENT

According to Dr. R. O. Slatyer, chairman of Australia's national committee for the Man and

75

the Biosphere program, numerous areas proposed by the states as "biosphere reserves" or "ecological reserves" have been cut down to eight for immediate certification to the United Nations agency, with 20 as the ultimate goal.

According to a recent circular of the UNESCO program, some 200 reserves have so far been proposed worldwide, but as of September 1976 only 57 had been accredited as fully qualifying. Of these, 28 were in the United States, ranging from Noatak National Arctic Range in Alaska to Big Bend National Park in Texas. Four biosphere reserves have been certified in Poland, nine in Iran, 11 in Britain and others in Norway, Thailand, Uruguay, Yugoslavia and Zaire. In each case a certificate is forwarded by UNESCO stating, in part: "This network of protected samples of the world's major ecosystem types is devoted to conservation of nature and scientific research in the service of man. It provides a standard against which can be measured the effects of man's impact on his environment." The latter sentence refers to the role to be played by the reserves in the Global Environmental Monitoring System that is slowly evolving as a sequel to the United Nations conference on the human environment held in Stockholm in 1972.

Establishment of the biosphere reserves is being supported financially by the United Nations Environmental Program. Once they are certified, the local government is to provide facilities for scientists, including those from abroad, to seek clues to trends caused by human activity. Such a clue, for example, could be the decline of some species vulnerable to one form of pollution. It is also envisioned that the sites be instrumented to monitor environmental changes directly. The reserves must be large enough—including, if necessary, a surrounding buffer zone—so that the native species can thrive and remain numerous enough to preserve their genetic diversity. In this way, as well, they can evolve.

EVOLUTIONARY RESPONSIBILITY

That the reserves should provide the potential for continuing evolution has been promoted by Sir Otto Frankel, who headed Australia's

participation in the International Biological Program (I.B.P.). In his words, man, having acquired dominance of world ecology, has acquired "evolutionary responsibility." The same theme has been championed by R. L. Specht, who headed Australia's terrestrial conservation program in the I.B.P. He told a conference on Australia's ecological reserves that extinctions are just as much a part of nature as evolution. In trying to save an "endangered species," he said, decisions have to be made as to where the line should be drawn, financially and in terms of other commitments. Despite the emotional appeal of preservation, he said, rescuing large areas of natural habitat is more likely to have long-term value.

Dr. Peter Crowcroft, head of the Taronga Park Zoo in Sydney, puts it differently. In his view: "The animal community is never in a static state. It would be more exact, or less inexact, to speak of an unbalance of nature than of a balance of nature, where animals are concerned."

"Most species," he said recently, "are capable of wide and rapid fluctuations in numbers without permanent harm." What ensures their disappearance, he added, is loss of their habitat "to the extent that there is only room for populations too small to be self-perpetuating."

Considerable attention has focused here on defining the size of reserve needed for perpetuation of a particular ecosystem, or complex of plant and animal inhabitants. If it includes large, roving species such as wolves, bears and red kangaroos, the area clearly must be large to hold a self-perpetuating population. As has been pointed out by Dr. Slatyer, it also must take into account the population density of the most rare species. He cited as test cases the islands that remain from the once contiguous land bridge to New Guinea. They became isolated after the sea rose at the end of the last Ice Age, some 10,000 years ago.

On islands with an area of 5,000 square kilometers (1,800 square miles), 51 percent of the bird species that, it is assumed, once lived there (since they continue to live on the nearby mainland) have died out. Where the island area is only 100 square kilometers (36 square miles), 92 percent are gone.

Barrow Island, a land bridge relic with an area of 200 square kilometers, no longer has any

A Matsheie's tree-kangaroo is a rain forest inhabitant of Australia.

A red kangaroo in Tidbinbilla Refuge.

red kangaroos. Yet 2,000 euros, another big kangaroo species, remain.

In terms of population densities on the mainland it has been estimated that the island was once the home of 200 to 300 red kangaroos, but the area was not sufficient for them to be self-sustaining, whereas for the euros it met the requirement. Dr. Slatyer noted that a number of the recently created Australian reserves are probably too small by the same criteria. Unstable situations currently exist on some offshore islands, as noted by Dr. Ride. The banded hare wallaby is plentiful on islands in Shark Bay, off Western Australia, but from past experience on the mainland, it is clear that those marsupials would be doomed by any introduction of foxes or livestock.

SURVIVAL REQUIREMENTS

Conservationists can also be misled, according to Dr. Ride, by assumptions of "rarity." A species may seem rare because it is sparse, even though its numbers, over an extensive area, are large. It may be good at hiding or naturalists may not know where to look for it, leading to false estimates of its true rarity.

Another problem is defining what a species needs to survive. The wombat, a marsupial that lives in forest burrows but at night ventures into clearings to eat grass, apparently cannot tolerate heat. Radio tracking and remote temperature measurements have shown that it must spend its days in a cool highland burrow. Its continued existence therefore depends on preservation of a special combination of circumstances. In a recent study, Drs. John H. Calaby and C. H. Tyndale-Biscoe concluded that in general at least 1,000 breeding individuals are needed to retain the genetic diversity and resilience of a species.

Australia has by far the most diverse species of marsupials in the world—those mammals that, as a rule, raise their young in a pouch. The continent is also said to have the most species of reptile. Its biologists are concerned that sufficient reserves be set aside to retain this diversity, despite competing agricultural and forestry interests. The rich genetic heritage of this continent, unlike others that are less isolated, has few parallels elsewhere.

The opportunities that it offers future generations for a deeper understanding of evolution and of diverse forms of life's functional processes must, they feel, be preserved. Many of those interviewed felt the new, relatively conservative

The Australian outback is home for kangaroos. This one has been chased and caught by its powerful tail for a closer look.

government now in office is less sympathetic to this than its Laborite predecessor. The latter created a department for the environment that has, it seems, been reduced to impotence. While biologists here and elsewhere believe evolution and extinction go hand in hand, they look to the new UNESCO system of biosphere reserves and its Australian analogues as a way to avoid needless, manmade extinctions. "In these reserves," wrote Dr. Ride, "representative sections of each major kind of natural Australian community shall be protected, and where necessary managed in order to retain, in all different parts of this country, examples of what nature herself had created over the millions of years during which the Australian environment has evolved."

"If we can succeed in doing this," he continued, "we will ensure that no species of Australian mammal, animal, or plant shall ever become totally extinct through man's actions." This is the goal of all UNESCO biosphere reserves.

Barry Commoner:
Scientist-at-large

by Alan Anderson Jr.

Some months ago, I accompanied Barry Commoner, the 58-year-old biologist, as he flew from his base in St. Louis to the far boggy north of Michigan to address a mixed group of auto workers, environmentalists and community groups at a United Automobile Workers lodge. Two weeks earlier, when he called to suggest the trip, he had warned me he was "really going to shake them up": He was going to give labor the radical message that the capitalist system was not working.

Arriving at the lodge, Commoner did not look like a radical. His suit and tie, neatly-cropped gray hair and thick-rimmed glasses more closely represented his title of professor of environmental science at Washington University. His only symbols of protest were subtle—his shirt was cotton and his suit wool, both materials that require less energy to manufacture than nylon or dacron. "It's a little gesture," he said. On this particular occasion, he had been invited to "bridge the gap" between those who would clean up the environment and those who fear such clean-ups will cost them jobs. Before he arrived, David Brower (Friends of the Earth) had spoken for the environment,

Leonard Woodcock (U.A.W.) for labor, and labor had remained suspicious.

"I understand that I'm supposed to make peace," Commoner began, his voice still echoing the tones of his native Brooklyn. "I'll do my best, but it may result in some very real pain. The thing I want to remind you about is that the earth is the home of the whooping crane, but it's also the home of people. I happen to think that people are more important than whooping cranes. But the point about whooping cranes is that when they die, or eagles die, their death is a symbol of something very wrong with the ecosystem." The practice of basing industrial decisions on profits has led capitalism into a number of "anti-social" production habits, he went on to say, notably less reliance on labor and a dependence on the lavish use of energy and capital resources. As a result, he pointed out, we are now short on both capital and energy and long on labor—ailments that can be cured only by basic readjustments in our economic system. "There are deep, inherent flaws in the capitalistic system," he said. "But there has been no way proposed to reform it that doesn't actually make the system worse."

"I've heard Barry do this before," a woman who works for the Urban Environmental Coalition in Washington, D.C., said after his speech. "We're going to see if we can get him to talk about socialism in a study group this evening."

THERE ARE DISSENTERS

As much as Commoner appeals to those concerned about the environment and to the young, he is openly disliked by industry, mistrusted by labor and regarded with a mixture of envy and outrage by his peers. Unique as a scientist, he is able with hip rhetoric and moral fervor, a combination of science and showmanship, to tread the narrow path between the Establishment and the counterculture in this country, assaulting targets both within and beyond his fields of expertise. But his greatest talent—synthesizing and translating many fields of science for laymen—often makes him susceptible to the charge of academic trespassing. His reputation as a communicator was established with the success of his second book, "The Closing Circle," which criticized modern technology as wasteful in its uses of both energy and raw materials and which, to date, has sold more than 300,000 copies in 18 languages. That reputation was further reinforced with the release of his third book, "The Poverty of Power," about two-thirds of which was excerpted by The New Yorker in February 1976. In the past year Commoner has given more than 40 invited lectures, and has testified before Congressional committees six times. He produces a monthly column for the journal Hospital Practice, and writes frequently for other publications.

"I hope everyone is aware that there's a philosophy behind what I do," he said during our airplane ride to Michigan. "Most of the major political questions of the day—agriculture, transportation, energy, product safety—have a pretty heavy scientific content and each is connected closely with the economic system. So in dealing with industrial society today, you have to consider a large number of scientific questions.

"The second point is that if you ask for a decision on any of these questions, it always gets

to be a value judgment. One example I often use is the elm trees and the robins. In the 1960's, when it was shown that spraying elm trees with DDT was killing robins, the public had to decide which it wanted: elm trees or robins. Or take domestic oil reserves. The decision that has to be made is whether we are willing to pay the price of extracting the oil, and how it is going to be paid for. Now those are political questions. If we are going to have a democratic decision, there is simply no way but to supply the people with enough information to vote on the issues." Commoner spends much of his time trying to do just that.

"Let's look at what a nuclear power plant really does," he said to those assembled at the U.A.W. lodge. "It boils water. The most efficient temperature at which to do this is about 1,000 degrees. The temperature of a nuclear core is somewhere between 1 million and 10 million degrees. You have to better match the energy source with the task at hand." There were a few soft whistles from the audience, and Commoner looked around the room. "I could kill a fly on that wall by walking up to it with a flyswatter. That's an example of a good thermodynamic mix," he said. "I could also kill the fly with a cannon—a bad mix. That would get the job done, but it would do a lot of damage, too, and waste a lot of energy." He went on to delineate the social costs of such mismatches between modern technology's tasks and the energy used to accomplish them (unemployment, waste, high energy prices) and to propose that solar-energy absorbers are far better suited to boiling water.

HIS ECONOMICS CREDENTIALS

Frequently reminded of his lack of credentials in economic theory, Commoner remains unabashed. "I've never had a course in economics," he says, "but an economist is not a biologist either, and one of us has to step across the line to get at the big issues." He was thrilled when recently at a symposium at New York University he turned midway through his analysis of productivity to ask Nobel-prizewinning economist Vassily Leontief, seated behind him, how his economics were holding up and Leontief replied, "A-plus," his thumb and forefinger joined in a circle. Other economists are divided. Peter Passell of Columbia University, in a New York Times review of "The Poverty of Power," liked Commoner's analyses of production and energy but disapproved of his economics; Robert Lekachman of the City University of New York praised the whole book in The Washington Post. Commoner's economics are similar to a politician's or a journalist's: He studies what pertains to his theme and more or less skips the rest. There is little mention in "The Poverty of Power," for example, of money supply, inflation, consumer demand or other classic variables. His economic knowledge came from pulling books from the shelves of the university library and bouncing ideas off economist colleagues at lunch. His contribution is not to erect a new economic philosophy but to point out a powerful and worrisome trend: the making of economic decisions for the public by private interests.

Commoner's predilection for searching out the facts began when he was a schoolboy in Brooklyn, where his father, a Russian immigrant, worked as a tailor. (The boy was so shy that he had to take corrective speech in high school.) His uncle gave him a microscope which he used to analyze tiny bits of life he found in city parks. A born tinkerer, he converted an automobile headlight into an illuminator for his microscope, with which he was then able to demonstrate that electrical current stops the movements of paramecia.

From such modest beginnings he went on to Columbia ("This may sound easy today, but in the 1930's it was not simple for a Jewish boy from Brooklyn to get into Columbia"), then Harvard (a Ph.D. at age 23). During World War II he joined the Navy, where, ironically, he devised a way of spraying DDT from a fighter plane to combat tick-borne fever on South Pacific islands. In 1947, he joined the staff of St. Louis's Washington University, where his research work began with tobacco mosaic viruses and moved in the mid-50's to pioneering studies of free radicals, molecules or groups of molecules with unpaired electrons. (He shares a patent for a method of searching for oil by means of free radicals.) With the advent of the fallout issue, he began a slow turn toward matters of social scope.

THE ADISCIPLINARY CENTER

Today, Commoner is director of Washington University's Center for the Biology of Natural Systems, which he founded in 1965 with a $4.5 million grant from the now-defunct Bureau of State Services. The initial center staff was drawn from the fields of biology, biochemistry, biophysics, psychology, anthropology, mathematics, medicine and sanitary engineering. In the last few years, the already broad definition of "natural systems" has grown still broader and representatives of sociology and economics have been added. Center investigations have ranged a wide spectrum: from lead poisoning in the ghetto and urban dog behavior to the economics of organic versus conventional farming and the pollution of rivers in the Corn Belt by fertilizer leaching.

"We are not so much interdisciplinary as *a*disciplinary," says Commoner. "Our work is organized around specific problems of society. We allow the problem to determine the pattern of work. That is the most radical thing about the center, and I think it's largely responsible for our success. It's really a question of whom you're working for. If you do discipline-oriented work, you're working for your peers. If you do task-oriented work you're doing it for the public—the ones who have the problems. The result is that our work is far more likely to point out the issues and possible answers than that of people who are pursuing one discipline."

Work is an obsession with Commoner. His only hobbies are fiction (he reads constantly on airplanes—Doctorow, Bellow and "mind-dulling" spy thrillers) and an occasional movie ("real good old-fashioned movies that have a message, like 'All the President's Men'"). Sometimes on weekends he and his wife, psychologist Gloria Gordon, escape to their 450-acre farm 90 miles south of St. Louis, where a neighbor grazes 100 head of cattle. Their daughter is a textile restorer living on the East Coast, and their son is studying math at Cambridge.

If Commoner is asked to pinpoint his mission—is it journalism? science? entertainment?—he will say that he is in the "information movement," a movement he regards as indispensable. It is his contention that we cannot entrust the nation's leaders with the job

Tampering with the environment: Commoner (right), during World War II, when he headed a Navy project to combat tick-borne fever on Pacific Islands by spraying DDT from fighter planes.

of getting all the facts to the public. "Despite conventional wisdom," he has written, "the Government's scientific advice has not been the essential input to decision-making. Quite the contrary; as in the ozone and nuclear-power issues, the decision-makers have usually been influenced—as, in a democracy, they should be—chiefly by the public." As examples of issues raised by the public in default of government action, Commoner lists the hazards of radioactive fallout, the detrimental effects of synthetic pesticides, the lack of energy alternatives and the danger of the huge reservoir of nerve gas near the Denver airport.

"During the fallout period of the 1950's, I was making speeches in every church and hall in St. Louis, describing the facts of atmospheric testing. Invariably, partway through the talk, someone would jump up and say, 'You mean we've been poisoning our neighbors' wells!' Exactly. Once they had the facts, their consciences began to function. At the same time, the Atomic Energy Commission and the Administration simply were avoiding the debate. Finally we were able to crack through this silence with facts about strontium 90 in babies' teeth, and once that happened the public was able to decide. The minute we could show there was a hazard, they [the A.E.C.] gave in."

DECISIONS BEYOND THE PUBLIC

A primary thesis of "The Closing Circle" is that much of the environmental crisis was due not to overpopulation (here Commoner broke with Paul Ehrlich and zero-growth advocates) but to postwar technological decisions in which the public had had no input. These decisions—to replace cotton and wool by synthetic fibers, soap by detergents, manure by artificial fertilizers, railroads by trucks—called for heavy use of petroleum stocks and energy, and produced goods that do not degrade when thrown away. Commoner described the resulting depletion of natural resources and pollution as "subsidized by society." When the book was published in 1971, with the economy booming and unemployment down around 4 percent, it predicted that "a full-blown crisis in the ecosystem" was "the signal of an emerging crisis in the economic system." The economic crisis came two years later, triggered by the Arab oil embargo.

In "The Poverty of Power," Commoner again proposed that if the public had access to all the facts, many decisions about energy and technology might be made differently. For example, he argued, using industry and Government figures, there is much more petroleum in this country than the public believes. The figure usually publicized, says Commoner, represents the supply obtainable at current prices; a far greater amount is available if we are willing to pay the extra $39 per barrel or so that it would cost. His point is that the public should be offered that option.

Similarly, he points out, the drawbacks in the conversion of coal to gas or liquid have not been spelled out to the public. For one thing, both converted coal and shale oil contain well-known carcinogens. For another, the gasification of coal means a 92 percent reduction in its capital productivity compared with its productivity when burned directly. He documents arguments that nuclear power is an "illusory solution" to the energy crisis because uranium resources will run out even before oil, and that the breeder reactor has been all but removed from consideration by its own huge capital costs. His solution? "Most recent studies conclude that midway through the 21st century, or even somewhat sooner, we could obtain all, or nearly all, our energy from the sun." Until then, he says, there is enough oil—if we stop wasting it.

"The Poverty of Power" makes it clear that, because the production system depends on the environment, it should be designed to be compatible with it. In fact, argues Commoner,

the production system insults the environment—and the human beings who are part of it—and these insults derive from conscious decisions by private enterprise.

For example, Detroit decided after World War II to produce high-compression engines, which yielded the nitrogen oxides that interact with sunlight to make smog. Earlier engines were incapable of doing this. Again, the petrochemical industry decided to make plastics out of vinyl chloride, a gas that causes cancer. Commoner attempts to show how these and other production decisions took wasteful advantage of cheap energy and raw materials, and were based solely upon the products' profitability. He concludes "The Poverty of Power" by questioning the profit motive as the dominant criterion in decision-making, when it is clearly in conflict with—and works toward the great detriment of—social goals.

STEPPING ON TOES

In this bold analysis, which Commoner wrote in longhand between July 1 and Nov. 15, 1975, he admittedly steps on a lot of toes ("I know, we barge into agronomy, we barge into economics"). He and the center are unpopular with the fertilizer, automobile and petrochemical industries—and even with their own university: The chancellor tried unsuccessfully to disband the center in January 1976 and incorporate it into the biology department. No reason was singled out, although some university people feel Commoner dominates the center to an unhealthy degree and tends to be, in the words of one figure involved in the brief struggle, "too eager to prove his case." He also lost points with his colleagues when he was described in Time magazine as responsible for a technique for screening chemical carcinogens by means of bacteria. The method was actually worked out by Dr. Bruce Ames of the University of California at Berkeley (Commoner says he tried to make this clear) and is being applied at the center.

Florence Moog, head of Washington University's biology department, objects strenuously to the publicity Commoner receives ("he seems to have a direct pipeline to the press") and to his mode of operation, but concedes that he is "very brilliant" and has an "exceptional capacity for digesting a vast amount of information."

Commoner was chagrined by the magazine misattribution, but shrugs it off as "one of the hazards of doing your social duty as a scientist." Social duty is another recurring theme with him; he likes to say his job is to "go in hot pursuit of the truth," wherever that may lead. For that purpose, with anthropologist Margaret Mead and others, he founded the Scientists' Institute for Public Information in 1963. S.I.P.I. now has a staff of 18, a budget of $300,000 and two magazines: Environment and Job Health News Service. "The scientific community," he has written, "custodian of what we know about the earth and its inhabitants, has important responsibilities toward the resolution of the grave economic, social and political conflicts that surround these issues."

SCIENCE FOR THE LAYMAN

This is where Commoner parts philosophic ways with many of his colleagues who agree with the "custodian" metaphor but demur when it comes to social conflict. Traditionalists say that the place for science information is in scientific journals; if there is to be debate, let it be not so much a "hot pursuit" as a reasoned dialogue, preferably in the pages of those journals. Alvin Weinberg, former director of the Oak Ridge National Laboratory, is typical of much of the traditional science community in his criticism: "Barry Commoner speaks rather irresponsibly in The New Yorker articles. He has no business saying these things in such an ex cathedra manner."

Philip Handler, president of the National Academy of Sciences (a body comprising supposedly the greatest living American scientists, to which Commoner has not been elected), charges that Commoner often begins with a bias: "Mr. Commoner does not have the facts and the analysis of facts separated. Therefore, he is coloring his facts." Says Commoner: "I would absolutely object to that characterization. I use an almost didactic and rigorous pattern. I always lay out the facts on which an analysis is based. Take the analysis of the Rasmussen

report [a report by M.I.T. professor Norman Rasmussen which showed the chances of a harmful accident at a nuclear power plant to be statistically remote]. I laid out the facts in a pretty fair way, I think. I then began an interpretation, which in that case was very narrowly aimed: Talking about the probability of a large nuclear accident is very different than the chance of an accident walking across the street. The latter is a conscious decision; with nuclear power you don't have a choice. And this question must be decided by the public."

Some of this, of course, is rhetoric. There are some themes dominant enough in Commoner's work to be called biases; his dislike of nuclear power dates back to the fallout days. He advocates organic farming, the efficient use of energy and a halt to the release of pollutants into the atmosphere. When N.A.S. president Handler suggests he is "coloring his facts," Handler is referring partly to the way Commoner selects examples to support his own arguments. But these arguments stem from ultimately fair questions that are looked at in a scientific way and are answered factually. He can be accused of "making a case," but not of fabricating answers.

Margaret Mead has been an ideological ally of Commoner's since the fallout days. Although she does not always agree with him on specifics, she defends his approach with characteristic vigor: "He has stood throughout for the complete presentation of facts so that citizens can make up their own minds. The science information movement, which Barry Commoner conceived of and built, holds that it is the scientist's duty to analyze facts and present them so the citizen can understand. I think it's insulting to laymen to ask them to make up their minds about something they don't understand." She defends with equal vigor the breadth of the projects undertaken by the center: "He's perfectly competent to lead the research his center does. He is interested in the whole and not in little fragmented bits. The scientist who attempts to hide behind his little bit of competency will always resent this. There are those who are sheltered in their narrow expertise, and those who will take responsibility for the well-being of the planet."

The clamor for access to "true facts" has in recent years made quite a racket in the halls of science. The bloodiest battle has been over nuclear power, whose engineers, physicists and vendors have seen themselves as forces of reason and servants of necessity embattled with bands of irrational Luddites hurling brickbats and half-truths. Lesser disputes have raged over SST's, food dyes, fluorocarbons in aerosols, cyclamates, the use of pesticides in the United States, of herbicides in Vietnam, pipelines in Alaska, fireproofing in children's pajamas.

VALUES IMPOSED ON SCIENCE

To traditional scientists, the most distasteful aspect of these debates has been the imposition of social values on pure science. The unease and frustration engendered by such clashes has created a strong desire for some insulating mechanism to "separate value judgments from scientific questions," in the words of one scientist. The boldest attempt to do this is the idea of the "science court," a concept that has created controversy in its own right. Like a judicial court, a science court would provide the mechanism for an adversary hearing, in which proponents of differing scientific positions would argue their cases before a panel of equally disinterested but well-informed scientist/judges, who would then present their decision to the public. Dr. Arthur Kantrowitz, chairman of Avco Everett Research Laboratory in Massachusetts, has been promoting the idea for 10 years as the best way to achieve "democratic control of technology." "But it's an untried procedure," he says, "and there is sort of a chicken-and-egg problem: Until the science court is established, you can't get people to take it seriously."

Commoner opposes the court idea on the grounds that its decisions might approach the status of "real truth," and that, thus, the system could grow authoritarian. The National Academy of Sciences is also chary of the scheme, and its support is essential. "The adversary procedure was invented to find out who wins," says N.A.S. president Handler, "and I've never heard of anyone winning in science. This must be absolutely free of the chance of some Perry Mason-like figure getting a chemical 'off the hook.'" The academy has its own method of

getting at "the truth"—some 550 committees composed of unpaid experts—a method Handler prefers to Commoner's approaching the public directly.

Simon Ramo, vice-chairman of the vast T.R.W. aerospace conglomerate, has chaired an advisory group which helped facilitate the recent appointment of a Presidential science advisor. While Ramo says there are "some details" on which he disagrees with the direct approach, he defends public discussion. "A lot of people are concerned about Barry Commoner because he writes so much and he's so articulate. All in all, I take a generous view about people writing and speaking. If he takes a broad topic, I think I can understand that he can't be an expert on all the things he raises. If he's wrong about certain things, let people write and speak out and say he's wrong. Scientific debate should be more like a tennis court than a science court. You hit the ball back and forth, and call attention to your opponent's errors by calling the ball out."

Commoner cites recent surveys by Louis Harris and by the National Science Foundation, both of which indicate that scientists and their work are held in high esteem by the public. "Science has a well-justified reputation for approaching the truth," he says. "Why? Not because we are any more truthful than anyone else, but because we have a tradition of making our mistakes in public—that's called publication. What stands in the way of a scientist faking his results is the certain knowledge that he's going to be checked. And you can bet that a lot of the the stuff I write is checked more carefully than most."

He also says that his heavy schedule of speaking and writing is "part of my job." Most of the center's funding (the budget is about $1 million a year) now comes from N.S.F.-R.A.N.N.—the National Science Foundation's Research Applied to National Needs. By contract, Commoner has to demonstrate the "user" value of his work. "I have to show who needs answers to what I propose to study, and how I propose to inform them."

Commoner's audiences often come away with a mixture of satisfaction and frustration. They have heard someone "lay out the facts" for them, as promised, but those facts seem to raise more questions. He is trying out the word "socialism" because it implies a more public form of decision-making than we now have, but he readily admits he has no idea of how such a system would work, or what would replace the profit motive as an economic incentive. "If I did," he says, "I'd be the most famous person in the world. It's a seductive but very dangerous idea to think that anyone who analyzes a problem ought to solve it. Nobody has tried to combine the egalitarian benefits of socialism with the freedom of the capitalistic system. The reason is that nobody has shown the need for it. We've tried to show the need; now I hope someone will come up with a solution."

Like a politician or a film star, Commoner flourishes in the limelight and thrives on debate. When asked if he thinks of himself as a star, he gives something like an embarrassed smile and says, "You'll have to decide that yourself." And indeed it is Commoner's willingness to come this close to stardom that gets him in trouble with some of his peers. He doesn't talk or behave like an orthodox scientist, and yet he says he is one. He gets more attention than other scientists who may think they work harder, or do "purer" science or bring less "coloring" to their facts. Commoner knows all of this; he is nothing if not perceptive. But he decided consciously and long ago to accept the risks along with the freedoms of being a maverick.

Scientists Struggle to Preserve Hawaii's Strange Species

by Walter Sullivan

There are inchworms that have turned the tables in Honolulu on their natural enemies. The only known predatory members of the large order of Lepidoptera, or moths and butterflies, they snatch spiders and devour them. Birds of the honey-creeper family have developed bills said to be more specialized and diverse than those of any other bird family.

Such evolutionary wonders derive from the remoteness of the islands, the most isolated archipelago on earth. The islands have become a prime laboratory for research into evolutionary processes. Nevertheless, many of the most remarkable species have become extinct, and the battle to save the rest is at a critical stage. Under suspicion as the causes of extinction are introduced diseases, displacement by foreign species and loss of habitat because of real estate development, agriculture and lumbering or because of defoliation by sheep, pigs and particularly goats.

In some cases the tide has at least temporarily been turned. Fencing off a plot in a region made barren by goats has led to the appearance of a previously unknown bean. In sanctuaries on the islands of Maui and Hawaii the néné or Hawaiian goose, once down to a free-living population of

less than 40, has risen to about 1,000 and has again begun breeding in the wild.

Nevertheless, according to wildlife specialists here, more birds have become extinct since the islands were discovered in the 18th century than in any other biological province on earth. Because this is the world's most isolated archipelago, its inhabitants were highly vulnerable to intrusion. Of 161 bird species that have vanished since 1600, 149 lived on islands, 25 of them in Hawaii. Only a dozen were continental. Of the 38 native Hawaiian land and freshwater species that remain, 25 are endangered, more than in the other 49 states combined. Most of the survivors live at high elevations on Hawaii's volcanic mountains.

A major blow to the native birds was the activity of the Hui Manu movement that, early in this century, sought to make Hawaii "the aviary of the world." Some 150 foreign species have been released and 32 have become established, including such residents of New York suburbs as the cardinal.

According to Sandra van Riper, visitors to Waikiki never see a native bird, and Diamond Head, the famous landmark overlooking Waikiki, is dominated by African finches. She and her hus-

88

Diamond Head, the famous landmark overlooking Waikiki is dominated by African finches, birds not native to Hawaii.

band, Dr. Charles van Riper, are assessing the role of malaria in the extinctions of native species and have found, so far, that foreign species, such as the Japanese white-eye, do seem to be more resistant. Dr. Donald Gardner of the National Park Service is studying diseases that attack native vegetation.

Some specialists believe the chief threat, apart from disease, is the removal of vegetation for which the native birds are specially adapted. One such case has led to a particular emotional confrontation between hunters and conservationists. This is the fate of the palila, a honeycreeper whose bill is designed to open the pods of the mamane tree.

Sheep, running wild on the slopes of Mauna Kea, Hawaii's highest mountain, are threating to crop that tree to extinction. Fencing off part of the area, some fear, would have little effect; the mouphlan sheep, imported from Asia, can jump most fences. Their number has already reached 400, according to Donald W. Reeser, management biologist of the National Park Service. A suit by the Sierra Club under the Endangered Species Act seeks to eliminate the sheep entirely, a move bitterly opposed by hunters. Mr. Reeser has been battling wild goats in the 80,000 acres of Hawaii Volcanoes National Park since the 1960's, and it was his efforts to cope with them by fencing a barren patch of land on the slopes of Kilauea Volcano that led to discovery of the long-lost bean.

After two years, reported a botanist, Dr. Harold St. John of the Bernice P. Bishop-Museum in Honolulu, the patch "from even a mile away was a conspicuous green, oblong spot on a light colored bare landscape." Wild goats, however, "were lined up at the fence, like cattle at a feed rack, stretching to reach a bit of an edible plant." Half the patch was covered with a previously unknown species of Canavaia.

By 1971, about 100,000 goats had been hunted down in the park, but this only seemed to invigorate an otherwise famished population. In the last four years, however, by fencing off and clearing sections of the national park one by one, the population has been cut from 15,000 to about 400. Several species of predatory inchworm have been identified and studied by Steven L. Montgomery of the University of Hawaii. When an insect touches the posterior end the front end of the worm whips back in a 12th of a second and grabs its victim with six legs clustered around the head. The inchworms become moths as peace-loving as any other Lepidoptera.

Mr. Reeser and his colleague, James K. Baker, in a plea for preservation of the remarkable diversity of Hawaiian plants and animals, have termed them "treasures that belong not only to the people of Hawaii and the rest of the United States; they belong to all the world."

Restrictions Are Impeding the Use of Solar Devices

by Gladwin Hill

Last year Paul Keplinger, a heating and plumbing contractor, built a solar energy collector in his front yard in Arlington, Va., a not unattractive cedar-framed arrangement of translucent panels 10 feet long, 6 feet wide and 5 feet high. It was hardly completed when Arlington County zoning officials told him that it was too close to the street and he would have to move it. Mr. Keplinger replied that meeting the zoning requirement would put the collector in partial shade and decrease its effectiveness by as much as half. He suggested that leeway should be given to solar devices, considering the national interest in conserving energy. The matter is still being argued. Mr. Keplinger's troubles are indicative of a thicket of problems that loom, apart from technology, as potential obstructions to early and easy exploitation of the virtually unlimited reservoir of power latent in sunlight. These obstructions include zoning laws, building codes, a lack of specific regulation governing solar devices and a legal no-man's-land involving people's access to sunshine unobstructed by neighboring buildings or trees.

A nationwide check by The New York Times indicates that head-on collisions with the law, as in the Arlington, Va., case, have been remarkably few so far. But many authorities believe the reason is simply that solar heating is still a relatively small scale innovation, with possibly a few hundred thousand installations nationwide, not that the potential conflicts do not exist.

From a technical standpoint, solar energy collectors for buildings are not a formidable problem. They are simple in construction, being essentially an arrangement for exposing a circulating liquid or air to sunshine, and are not inherently unsightly or unduly heavy. A typical household solar water-heating installation costs $1,000 to $2,000 and may pay for itself in 10 years or less. Home solar space-heating installations nearly always have to be coupled with conventional heating systems and may cost 10 times as much, which makes amortization within the lifetime of a typical home a marginal proposition.

Christopher Raphael, an architect in Reston, Va., tends to wax caustic about the attitude of Virginia officialdom toward solar heating. "I had to build my house with solar and also conventional heating," he says. "The state's not willing to let an owner accept the responsibility for a cold home."

In Pittsburgh, Pa., as in Arlington, zoning laws restrict the placement of solar equipment. Neill

Barker, manager of solar systems sales for P.P.G. Industries in Pittsburgh, was chagrined to find that a deed restriction on his residence forbade equipment on his roof, as well.

In Anaheim, Calif., a solar heating installation in one particular subdivision ran afoul of a beautification regulation forbidding any sort of roof-mounted equipment. The State Energy Commission persuaded officials that the restriction, antedating the energy crunch, should not be carried so far.

Even Florida, a state in the forefront of the solar energy movement, is not immune. Ron Yachobach, president of the Florida Solar Industry Association, said that building codes in Miami and Fort Lauderdale "are slowing the growth of solar energy because they're both too vague and too stringent, and most solar collectors don't pass their unreasonable tests."

Generally, however, the trend seems to have been on the side of liberality. Twenty years ago, the city of Coral Gables, Fla., banned solar water heaters on homes for esthetic reasons. The ban has

been relaxed and now home owners are saving hundreds of dollars a year on fuel bills. "On average," one resident said, "I only have to use my electric booster once a week for about 30 minutes."

"People who have trouble with local building standards," says Patrick McCuan, a Columbia, Md., builder who is putting up 12 solar-equipped homes, "are the ones who don't do enough advance research. Zoning and building regulations are no more stringent for solar-equipped homes than for regular homes if the builder shows officials exquisite plans. This is necessary because specific regulations just aren't there yet, so it's a learning and negotiating process for both parties.

At present this absence of explicit regulations on solar equipment is more of a problem than restrictions are.

"Inspectors don't know much about solar energy installations, and when the code doesn't have something, they're reluctant to give approval," said William Osborn, director of the Massachusetts Solar Energy Office.

A number of states are moving to fill the regula-

Eddie Chow hosing off a solar energy collector on the roof of his home in Anaheim, California. His installation ran afoul of a local beautification ordinance, and he had to seek help from the State Energy Commission to keep it.

Richard D'Aurio built a solar energy collector on his lawn but his neighbors thought it was ugly and the town in which he lives said it violates zoning statutes and ordered him to remove it.

tion gap. Oregon has nearly completed a protracted public-review process on a "uniform solar energy code." New York's Legislature has just enacted a "solar zoning" bill that sets out solar guidelines for the local planning and zoning authorities and asks their collaboration in anticipating solar needs through such devices as laying out streets in new developments to provide southern exposures for homes.

California's State Energy Commission has given $400,000 to the League of California Cities to develop a training program for local officials aimed particularly at eliminating any barriers to solar energy in building and zoning codes and land-use plans.

"Zoning laws can hinder or facilitate the deployment of solar systems within a community," observes Richard Maullin, commission chairman. "Zoning restrictions may constrain a developer's ability to vary lot sizes and setbacks as a means of maximizing solar use. On the other hand, zoning restrictions on height, location and use of buildings and plants may serve to protect access to sunlight for nearby property."

The California energy commission estimates that one-third of existing dwellings cannot accommodate solar energy collection apparatus because they are oriented poorly in relation to the sun (a southern exposure is best) or because they are shadowed. Common law gives a property owner sacrosanct "air rights" extending vertically upward from his property line, but that may assure only a very limited amount of sunshine. The extent to which sunlight or view may be blocked by new adjacent structures is a classic unresolved issue.

The problem of trees and shrubs that cast shadows on adjacent properties is particularly knotty, and is something of an environmental dilemma. Conservationists are strong for solar

92

energy, but they also are strong for urban trees and shrubs, which refine some air pollutants, produce oxygen and muffle noise.

"Many different proposals have been made for solar access legislation," says Gail Boyer Hayes, energy program director at the Environmental Law Institute in Washington, "but none of them deal comfortably with the problem of vegetation control. Restrictions on the species, height and placement of trees have the potential of becoming the most controversial issue in solar access legislation."

The Clamor to Poison the Coyote

by Hope Ryden

Conservationists always operate at a disadvantage. They must win every battle or the resource they are defending may be destroyed. Exploiters, on the other hand, can regard their defeats as mere setbacks to be challenged at a later date. It was predictable, therefore, that the 1972 Presidential order banning the use of poison to kill wild predators on the public domain would not be accepted by sheep ranchers as the final word.

In the West, most commercial sheep are pastured on Federal lands. The lands, and the wild predators that reside there, belong to every American. Yet Government traps, snares, bullets, fire and gas destroy these animals by the thousands on behalf of the sheep industry. And wool growers have never ceased clamoring to use poison again.

In 1975, they were successful in getting sodium cyanide exempted from the Government poison ban. Explosive traps, called M-44s, baited with that deadly substance were re-installed across millions of acres of public lands. The trap shoots poison into the mouth of any coyote, bobcat, bear or fox that investigates its scented tip. Now wool growers are agitating to regain use of sodium fluoroacetate, or "1080." This poison is most deadly to

coyotes, but it can kill any creature that feeds on corpses of its victims. Last month to dramatize their demand, Idaho sheepmen closed their private lands to hunters. Access would be denied, they said, until 1080 is again made legal.

One would suppose from this that wool growers had experienced economic hardship as a result of being denied 1080. Not so. Since 1974, the wool industry has realized higher profits than prior to the 1972 poison ban. It is true that fewer people raise sheep today. But this fact (in itself a boon to our over-grazed public rangeland) can hardly be blamed on the coyote. No real increase in predation has occurred. Moreover, since 1972, the Interior Department's Division of Animal Damage Control has killed more coyotes annually than while 1080 was still in use. In the past six years, Federal predator control expenditures have nearly doubled. As a result, between 1974 and 1977 the coyote population in the Great Basin and Mountain States has been reduced by 26 percent. Still the sheep industry is not satisfied. One wonders how much Government support this group can expect. The Wool Act of 1954 compensates sheep raisers whenever the support price for wool exceeds the national average. In addition, high

94

A coyote feeding on a lamb just after killing it.

import duties discourage foreign competition and raise consumer prices on wool. More Government subsidy is available in the form of cheap grazing leases on public lands. And, finally, a Government predator control program saves wool growers the cost of paying shepherds to guard their flocks.

In the Mountain States, herders are employed by only 16 percent of the ranchers running sheep on public lands. In the Great Basin States, only 24 percent hire shepherds to protect their livestock. Of sheep enterprises over 5,000 head operating on the public domain, only 8 percent have constructed lambing sheds to shelter new-born animals. Fewer still use guard dogs to ward off predators. And none seem willing to try non-lethal chemical repellents to discourage predation. Why bother when a responsive Government agency is

so willing to slaughter the nation's wild carnivores on their behalf?

In 1973, a panel of impartial scientists headed by Dr. Stanley A. Cain reported to the Interior Department that massive extermination of coyotes on public lands produced no economic benefit. The cost of such "prophylactic killing," the report states, far exceeded any realistic measure of the worth of the livestock preyed upon. Since only a few ranchers actually sustain heavy losses to coyotes, the scientists recommended the selective removal of marauders in response to complaints of damage. The Interior Department's Division of Animal Damage Control has chosen to ignore that advice. Body-count is still the order of the day. Last year's tally of dead coyotes (91,573) is viewed as this year's figure to best.

A trapper for the State of South Dakota and the federal government sets a trap for one of the most elusive animals on the continent, the coyote.

The impact of all this slaughter on whole ecosystems cannot even be guessed. Without coyotes to hold them in check, rodent populations become a problem. An environmental impact statement on predator control is long overdue. Instead, the Interior Department has prepared a 400-page option paper evaluating various approaches to coyote management. One piece of information contained in the paper will surprise taxpayers. Western sheep production is dominated by a handful of ranchers. Fifty-six percent of all sheep on public lands are owned by only 6 percent of Western ranchers. That means fewer than 3,000 individuals are the main beneficiaries of all the high-priced killing.

Tiny Town vs. Mining Giant

by Roger Neville Williams

Crested Butte sweeps its rocky ridge into the Colorado sky, high above the tiny hamlet which has taken its name. More than a butte, it is a 12,000-foot peak with a saw-toothed crest, a mini-Matterhorn challenged only by the supine Mount Emmons across the valley. The two mountains stand like bookends above the plain.

In between lies "The Butte," an ex-coal-mining town of 1,000 people, many from New York, California and Atlanta, who have trimmed out the old village in Disneyesque fuchsias and lavenders while adding stately Victorian buildings of their own. The entire town is a national historic site, almost entirely surrounded by National Forest land.

Three miles to the north is ex-Secretary of the Army Howard (Bo) Callaway's ski area, on the rim of the vast cirque edging the valley, where Bo and his brother built their condominium village. It is a tasteful, modern ski complex. Crested Butte and the ski mountain lie some 25 miles south of Aspen as the bullet flies, or 225 miles from Aspen by paved road and mountain pass.

The townsite is in a remote, enchanting setting at the end of the road in one of Colorado's last unspoiled valleys. Crested Butte lies at the end of the road of some people's lives as well, a last frontier now threatened by massive industrialism unlike anything Western Colorado has ever seen.

AMAX Inc., formerly American Metal Climax Inc., a giant mining corporation with more than $3 billion in assets, wants to construct a billion-dollar molybdenum mine two miles from town, and the urban refugees who have settled permanently in Crested Butte, and who see themselves as preservationists of the surrounding wilderness and of the historic landmark to which they have brought a renaissance, are not too happy about it.

When I drove up the 28 miles from the sleepy town of Gunnison, past the abandoned, lichen-covered homesteads, past the ranchers' proud turn-of-the-century homes, I was met at the plain of the North Valley by hundreds of poky Herefords coming down the road, prodded along by mounted cowboys in snow-covered chaps—a real cattle drive. This is the Old West, still mythic in its appeal, a tranquil region of range and mountains, relatively unscarred

The town of Crested Butte is located across from Mount Emmons in one of Colorado's last unspoiled valleys.

by its one compact Victorian mining town and scarcely marked by the recreation boom of the last 15 years.

□

In 1977, two miles from Main Street and 11,000 feet up on Mount Emmons (known to locals as "Red Lady Mountain"), prospectors for AMAX sank their core drills into what is thought to be the world's richest known deposit of molybdenum ore. Colorado has two other larger deposits, at Henderson near Berthoud Pass, and at Leadville, both being mined by AMAX's Climax Molybdenum Division, but neither ore body compares to that of Mount Emmons, which has been estimated to have a market value of $7 billion. Worldwide demand for the "gray gold" is rising at 7 percent a year, and Climax Molybdenum—which provides nearly half of the free world's supply—wants the Red Lady's moly.

In addition to the mine on the mountain, AMAX has proposed a mill and tailings pond (slime dump) of staggering proportions, all of it to be set down on surrounding National Forest and ranchland that has never known modern industrialization. The mine and mill would employ 1,300 people, cost as much as $1 billion, and require more than 2,000 construction workers during the building phase.

Of the 165 million tons of ore estimated, conservatively, to be inside Mount Emmons, 164 million tons of it would end up *outside* the mountain, on the surrounding land. There will be a tailings pond, a useless, caustic, chemical-laden sludge that's left after the moly concentrate has been extracted, covering 3,000 acres, 30 times the size of the town of Crested Butte. Gunnison County's total population, currently 11,000, would more than double over the next six years. Bumper stickers seen around town say: "DON'T CLIMAX IN THE BUTTE." All anyone can talk about in town is "the mine." Crested Butte is a town fighting to save itself.

□

So what is molybdenum, aside from being nearly unpronounceable? (It's been called "Molly Be Damned.") It is a silver-gray metallic element with a melting point of 4,730 degrees Fahrenheit that is used to alloy most steel products, to provide durability as well as resistance to corrosion and severe temperatures. It lightens steel, while hardening it, and it is found in everything from tools and engine blocks to steam turbines, jet engines, pipelines and nuclear reactors.

Molybdenum has been used in armor plate and cannon since 1894. In 1905, a German company discovered the world's largest deposit at Climax Station, near the town of Leadville, Colo. The company, Metallgesellschaft, changed its name to American Metal Co. when World War I came along, but that didn't prevent the War Minerals Board from seizing its stock and placing it under the trusteeship of Henry Morgenthau and Andrew Mellon. A rival company, Climax Molybdenum, was then formed to provide moly for Allied war instruments. By World War II, molybdenum was the country's No. 1 strategic metal, and troops were garrisoned near Climax to guard the mine and mill from possible Axis sabotage.

□

When AMAX officials came to Crested Butte a year ago to announce their find, the last thing they expected to encounter in a tiny, rural community was a phalanx of resident New York lawyers, who, along with just about everyone else in town, want to keep the Butte and its surroundings as they found it.

There is New York University Law School graduate Myles Rademan, the town planner who came to the Butte six years ago and says: "You just can't imagine the size of this project, its magnitude. When you have 200 or 300 million tons of tailings, you're talking about a retaining wall the size of the Aswan Dam or bigger. Areas under consideration for tailings dumps are among the most beloved and beautiful parcels of land in the world. What we're dealing with here is the disemboweling of the West."

There is Gil Hersch, a former New Yorker who now owns and publishes the weekly Crested Butte Chronicle. He told me: "I'm opposed to the mine. We're killing our planet. I'm interested in the major paradigms in the world right now, between the traditional capitalistic outlook of more-makes-better, as opposed to the New Age culture which says we've got to live in harmony with nature. AMAX came in here with the statement that they were going to make this the first in a new generation of mines. We asked,

99

The Crested Butte ski area is located three miles north of the town on the rim of the vast cirque that edges the valley.

'What is a new generation of mines?' Is it that it's just 15 years newer? Or is it the difference between the Industrial Revolution and the New Age Revolution?"

And there is County Commissioner David Leinsdorf, son of conductor Erich Leinsdorf. He was recently re-elected by playing David to the AMAX Goliath. He advocates a much smaller mining operation than AMAX envisions. A Columbia Law School graduate, ex-Nader Raider and former Justice Department attorney, he moved to the Butte in 1971 and became Colorado's youngest county commissioner in 1974, acceding to a post normally reserved for conservative, native-born ranchers.

The stakes are high: A $7 billion mineral deposit versus the priceless value of wilderness; technological progress and modern industrialization versus a ranching and resort economy; transient workers and "boom town" greed versus a tiny, stable community practicing, for the most part, a kind of voluntary simplicity.

Recreation provides Gunnison County's chief income, as a million and a half campers, hikers, fishermen, back-packers, hunters, jeepers, and skiers visit Gunnison National Forest each year, and one might expect the resort industry to lead the fight to preserve the integrity of this region of unparalleled natural beauty. But that's not the case.

"I wish they'd found their moly somewhere else, but I'm not going to lie down in front of any trucks to stop AMAX. It's not my style," said Bo Callaway in his office overlooking his ski lifts. "I support the free-enterprise system, and I'm not comfortable telling a mining company when and where they can mine. If AMAX goes the extra mile and operates openly, I'll continue to work with them."

As President Ford's campaign manager, before he resigned over the "Callaway Affair" and moved permanently to Crested Butte, Bo was the enemy of the area's environmentalists. He proposed a bigger ski area, on another mountain, and was accused by Myles Arber, the former editor of The Chronicle, NBC News and Senator Floyd Haskell, the Colorado Democrat, of using his Cabinet post in Washington to lever the required special-use permit from the Forest Service. He handed me a copy of Harper's with the July 1977 cover story absolving him of any wrongdoing, his sweet vindication.

□

Money is money, and $7 billion is sitting inside Mount Emmons. Already it has had its effect—just lying there: $75,000 condominiums now sell for $200,000 and Crested Butte lots, which brought $10,000 last year, are going for $25,000 today. Terry Hamlin, a former head of the ski patrol, ski-area marketing manager, real-estate agent and a Colorado native, is a case in point. The young, personable ex-ski bum is now manager of local affairs for AMAX. "A man on the way up," taunts Gil Hersch in The Chronicle. A shotgun blast shattered the windows of Mr. Hamlin's AMAX office in the fall of 1978, but he was away at the time.

Hamlin drove me up to the mine site on Mount Emmons, two miles above town on the Kebler Pass Road. Already, 135 men and women are employed here in "exploration," bearded young geologists examining core samples, miners cleaning out and shoring up the drifts (tunnels) of the abandoned Keystone Mine. The tunnels of the defunct lead-zinc mine happen to lie directly beneath the huge molybdenum ore body and are being used for exploratory core drilling.

In 1977, AMAX took the mine over from a Wyoming firm, fully aware that the mine's 100-year-old tailings dump had collapsed, polluting Coal Creek, Crested Butte's primary water supply. Under a Colorado Department of Health cease-and-desist order, AMAX had to spend $800,000 to clean up the previous mine owner's mess and stabilize the old tailings pond. Terry Hamlin handed me a colorful, 18-page AMAX brochure that promised to bring brown trout back to Coal Creek, concluding, "AMAX is committed to the people of Colorado to help maintain the air, water and natural environ-ment that make the good life here what it is."

"Next summer we're building a $2 million water-treatment plant up here to completely end the Keystone drainage contamination of Coal Creek," Hamlin explained. "In the past, at the historic stage of the development of the West, mining companies were not concerned with the environment. But AMAX is. We have to be." It was a theme I was to hear again and again from AMAX officials.

But no one on the Town Council has bought AMAX's expensive public-relations arguments. "It seems highly unlikely that the mine and the town can be compatible, or that recreation and mining can be compatible," says William Mitchell, 35, Crested Butte's Mayor and chairman of the Town Council. "I'm very anxious about it. This is a national playground and it belongs to everyone. These are public lands. Those of us who've consciously given up conveniences and have adopted a harder way of life as a trade-off for the natural wonders surrounding us have become trustees of a magnificent wilderness."

□

Mayor Mitchell—he goes by his last name alone—has been using his "situation," as he calls it, to attract attention to Crested Butte's plight. After a 1971 motorcycle accident badly burned his body, hands and face (now restored by plastic surgery) and a subsequent airplane crash fractured his spine, Mayor Mitchell is now a paraplegic—in what he calls "the least wheel-chair accessible town in America," Mayor Mitchell says, "I'm way too selfish for self-pity. If I hadn't got burned up or been in a crash, I would not have enough money to be Mayor of Crested Butte, since the job pays only $25 a month and I'm at it full time." Mayor Mitchell's dynamism, charisma and tough leadership qualities, combined with his ex-radio announcer's voice, keep one from noticing his "situation."

"He's a one-man media event," says Myles Rademan, noting the national publicity afforded Crested Butte under such headlines as "Paralyzed Mayor Battles Mine Owners." When President Carter invited Mayor Mitchell to Washington to advise the President's Committee on Employment of the Handicapped, the Mayor dropped by the Interior Department to see Secretary Cecil Andrus, a man who, he notes, launched his career stopping a molybdenum mine at White Cloud, Idaho.

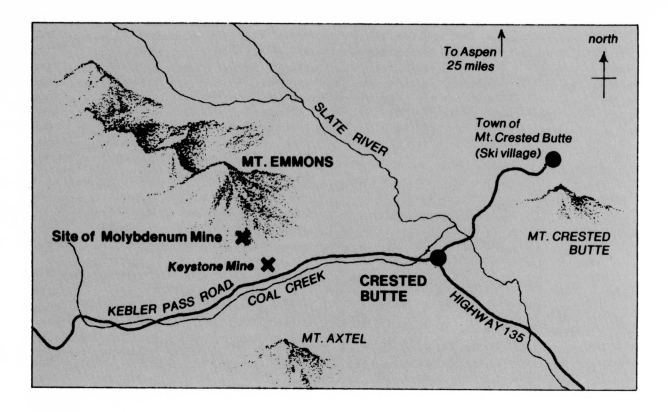

Mayor Mitchell confronts and confounds AMAX's media blitz (full-page ads in the local papers, a monthly sheet called The Moly News featuring columns by company executives, and a comic strip called "Miss Moly") with a tireless campaign of his own. In recent months, he has testified before the House Subcommittee of Mines and Mining, and has met with Paul Ehrlich, John Denver, Robert Redford, Jack Anderson and David Brower, president of Friends of the Earth. He once threatened to address an AMAX stockholders meeting, until AMAX chairman Pierre Gousseland and the then-president of the Climax Division, Jack Goth, invited him, along with Messrs. Leinsdorf, Callaway, Rademan and the Mayor of Gunnison, to visit the home office in Greenwich, Conn., last year.

"We can't give in to the divine right of ownership by mining companies," says Mayor Mitchell. "I'm fighting for every protection I can get for this valley. We can understand the value of Yellowstone and Yosemite—there are probably tons of mineral deposits in Yosemite and a decision was made in 1890 to preserve it—but what is the value of this valley? There have been more than 1,500 mining claims filed in this county this year. What they want to do is make Gunnison County one large open pit mine."

In November 1978, Mitchell was unhappy to see Colorado's environmentalist Senator, Floyd Haskell, a Democrat, defeated by William Armstrong, a Republican Congressman. Armstrong had been listed by Environmental Action as one of Congress's "Dirty Dozen" for his sorry environmental voting record while a Congressman in an adjoining district. But Mayor Mitchell is heartened by the position of Democratic Governor Richard Lamm, who recently counseled him: "[AMAX's] admission price to this community is going to be very high."

The result of Mayor Mitchell's activism has been an enormous amount of unprecedented cooperation between a giant corporation and a small community. Despite a unanimous vote by the Town Council on Feb. 20 to oppose the mine, the Mayor knows that it is hard to stop a billion-dollar-a-year corporation from doing what it wants. Messrs. Leinsdorf, Rademan and Hersch, lawyers all, also know that the Mining Law of 1872 guarantees mining companies access to all mineral deposits under public lands. And so they are fighting what the Mayor calls a "gentlemen's war."

I was unprepared for all this cooperation. I had

half expected to find a local Monkey Wrench Gang loose in the woods, pouring Karo syrup into bulldozers, dynamiting construction sites, rolling boulders down on workers' pickups. In fact, there has been some of that: a D-7 Caterpillar was riddled with rifle bullets, gas tanks have been sugared, and there was the shotgun blast through Terry Hamlin's window. A resident of Ohio Creek, an area threatened by the proposed tailings pond, said at a public hearing that if AMAX came to his valley, he would oppose the company by force, if necessary. AMAX, meanwhile, proudly points out in The Moly News that its security guards remain unarmed. So far.

In an attempt to anticipate problems before they happen, Messrs. Leinsdorf, Rademan, Hersch, Callaway and Mitchell and a talented young planner for Gunnison County, Jim Kuziak, are attempting an

untried and radical method of dealing with the massive mining operation. As Mr. Leinsdorf explained it: "I'm not as concerned with the environmental consequences as I am by the boom-town syndrome, the spin-off that comes from rapid industrialization and a high rate of growth: social and economic problems, unemployment, crime, drug abuse, alcoholism, child abuse, transient workers. So we're trying to see to it that the development that comes does not overwhelm us. It's called the Colorado Review Process, involving AMAX, the Forest Service, the Colorado Department of Natural Resources and the County."

The Colorado Review Process (C.R.P.) was devised for reviewing proposed ski-area development, but it has never been applied to mining. The process addresses such problems as housing, transportation, schools, sewage, air pollution and environmental im-

AMAX, Inc. wants to construct a billion-dollar molybdenum mine similar to their other installations.

The entire town of Crested Butte is a national historic site, almost entirely surrounded by National Forest land.

pact. The C.R.P. is the brainchild of Mr. Leinsdorf's old Columbia Law School classmate Harris Sherman, director of the State's Department of Natural Resources and a Lamm appointee. AMAX has signed an agreement with the Forest Servce, Gunnison County and Mr. Sherman's agency to participate in the C.R.P. It is an important breakthrough in a state that has probably been the least progressive in the West on questions of land use planning and mining controls.

Mr. Leinsdorf led the fight three years ago to have the Board of Commissioners adopt the county's Land Use Resolution, one of the most innovative documents of its kind in the country. He also insisted that AMAX and the Forest Service participate in the C.R.P. Mr. Leinsdorf has taken on corporate giants before: He was a co-author of the book "Citibank" for Ralph Nader, a critical look at the First National City Corporation. When the Crested Butte delega-

tion went to Greenwich to see AMAX's top executive officers, Mr. Leinsdorf opened the meeting by telling a story about his first job, in the Antitrust Division of the Justice Department.

"I was given a filing cabinet full of files on a possible restraint-of-trade and monopolistic-practices case," he began. "I was told to examine all the evidence and either bring charges against the company or close the case. That company was American Metal Climax. I was to see if a company which produced 70 percent of the molybdenum in the United States was engaging in unfair practices." As the executives tensed up, Mr. Leinsdorf laughed and added, "I found none and closed the case. But now, 10 years later, 2,000 miles west of New York City, I'm involved with AMAX once again. Which just goes to show, there really is no running away."

Mr. Leinsdorf believes that it is possible to make the Mount Emmons mine a model project. He sees

In addition to the mine on the mountain, AMAX, Inc. has proposed a mill and tailings pond (slime dump) of staggering proportions. This is the proposed tailings site at Carbon Creek.

the joint public review as an opportunity to develop a new process that could apply to the whole spectrum of natural resource development throughout the West: uranium, copper, shale oil, gas and coal. In that spirit, AMAX has contributed $100,000 to the Gunnison County Planning Department, toward a computerized planning system. But Mr. Leinsdorf also told me, "If AMAX is not willing to build the mine on a scale and a schedule that maximizes the benefits to the county, I'm prepared to fight them and vote no on their zoning-change permit. The only way to keep the tourist and ranching economy from being overwhelmed is to do it on a smaller scale."

At the first C.R.P. hearing at Gunnison High School in September, 1978, every opinion on the mine was heard in the long, emotionally intense and crowded meeting. AMAX officials listened and answered questions very carefully. AMAX emphasized employment and taxes, while the local governments

pointed to environmental degradation, social upheaval and irreparable harm to the tourist industry.

Although public sentiment at that particular meeting was predominantly antimine, a few retired Crested Butte miners spoke in favor of it, remembering the prosperous coal-mining days of the 30's and 40's. One Gunnison businessman said Crested Butte had an "obligation to humanity to supply molybdenum to the world." A native Crested Butte resident stood up and shouted, "I can't eat scenery!"

At AMAX's Climax Molybdenum Division offices in Denver, I spoke with Stanley Dempsey, AMAX's vice president for external affairs. A jovial, thoughtful man of 39, Mr. Dempsey has taken on the burden of getting the mine built. He has spent a great deal of time in the Butte, where he is respected for his candor and sincerity.

"I think we're lucky to come into a place like

Two of the historic buildings in Crested Butte are the Old City Hall (left) and the old Slogar Building (above) which currently houses a restaurant.

Gunnison County, where both the traditional community and the new-life-style community are very sophisticated. Despite a strong heritage against a lot of government control, the county has had the political will to pass a strong Land Use Resolution. Few places have that on the book yet. I'd rather have a lot of sophisticated enemies to work with than a bunch of dumb friends."

Stan Dempsey was in charge of environmental controls during the construction of the gigantic Henderson molybdenum mine, a project now hailed in the mining world, and by conservationists as well, as the most environmentally sound mine and milling operation in the world. "I don't worry about the physical impact, I think we can handle that," he says. "The growth-management issue is the challenge. That's why we really believe in open planning. We realized that the public was not going to put up with the kind of mining that's been done in the past and we know that we have to solve the housing and transportation problems which are the principal constraints on mining."

AMAX, at Mr. Dempsey's suggestion, recently organized an educational "boom-town tour" for 36 town and county officials to see what was going on in other mining towns. AMAX picked up the entire tab—charter planes, hotels and meals—for the tour of six Colorado and Wyoming boom towns recently "impacted" by energy and mineral development, including the infamous Gillette, often described as an "aluminum ghetto." Myles Rademan called me when he got back; the trip had backfired somewhat. "What we saw out there was a real mind blower. It was frightening. We will have to take a quantum leap to avoid becoming like them."

Stan Dempsey does not try to defend AMAX's environmental record at its Climax operation near Leadville, America's second largest hard-rock mine. Nor does AMAX's public-relations director, Terry Fitzsimmons, who represents the third-generation of a Leadville mining family—his grandfather worked for Horace Tabor (of "The Ballad of Baby Doe" fame). Mr. Fitzsimmons said, "At Mount Emmons we plan to build a mine in keeping with the values of today. When Climax was built, the last thing miners or anyone else cared about was the environment."

□

In the Butte, it's the *first* thing anyone thinks about, and the residents continue to worry about their mountain's collapsing. "Subsidence," mining men call it, and people in the Butte don't want Mount Emmons to "subside," to crack and slip and cave in, as Bartlett Mountain did at AMAX's moly-ore mine at Climax.

One particularly concerned Butte citizen is Susan Cottingham, 30, the chairman of the legal committee of the High Country Citizens Alliance. This 140-person group devoted to preserving the quality of life in the mountains is challenging the idea that mining is the highest and best use of the land.

The mine is not unstoppable, she told me. "AMAX has been stopped from doing projects in Wyoming and Tennessee. If a lot of aggressive and very intelligent people can't stop it here, the West might as well roll over and die. The technological

dream of a 'model mine' belies human nature. I like a lot of the AMAX front men, but I don't believe their promises. I guess I'm an environmental paranoid. But then just look at their record."

Gil Hersch, down at his cramped Chronicle office, did exactly that. He reprinted a devastating report by some Tennesseans that carefully documented AMAX's coal, zinc, lead and copper operations in six states. "[AMAX has] a record of noncompliance with the law and of disregard for local citizens," the report stated. "Moreover, AMAX has shown a pattern of aloof resistance toward public officials, employees or ordinary citizens who have called upon the company to match its image with its performance."

In Colorado, AMAX has fought all "severance tax" proposals (taxing minerals as they leave the ground)—even though the money would be allocated to "impact mitigation," something to which AMAX says it is dedicated.

Chuck Malick, 28, is president of the High Country Citizens Alliance. The leather-worker and shopkeeper came with his wife to Colorado eight years ago as part of "the hippie dream." Standing by a crackling Ashley stove in his barn-wood paneled living room, I asked him: "This was a mining town. What's wrong with its becoming a mining town again?"

"We're not against mining," he explained. "The Keystone Mine operated here until 1975 and employed about 50 people. I'd welcome several small mines. Our concern is that the public should have something to say about the management of resources on public lands, especially in a fragile area like this. What the nation is being asked to do in the West, by mining and energy corporations like AMAX, is to sacrifice our national playgrounds. The old adage that there's always someplace else to go is just not true anymore."

Then Mr. Malick added, "All this environmental stuff is not just flowers and trees. You kill the recreational trade and I'm out of a job. And right now there is no unemployment here, but in mining boom towns the unemployment figures are around 11 or 12 percent. We don't need that."

The High Country Citizens Alliance has been successful in persuading the Forest Service, which has traditionally been overly cooperative with mining and timber interests, to consider a "no-action alternative" in its environmental-impact statement—that is, no mine. The Forest Service also listened to the Alliance's and to County Planner Kuziak's objection to AMAX's plan to build two 16-

A million and a half campers, hikers, fishermen, backpackers, hunters, jeepers and skiers visit Gunnison National Forest each year.

by-16-foot drifts (tunnels) to the ore body; it persuaded AMAX to put the tunnels "on hold" for now. It is the company's first setback.

Newly appointed Crested Butte Councilman Kirk Jones, a hard-drinking construction worker, claims that he represents 85 percent of the town's residents when he says, "I'm a consequence of the 60's and I'm tired of hearing about Leinsdorf's 'model mine.' I say to hell with the mine. Let the moly stay there; it could be a national reserve. Maybe in 30 years they'll have the technology to take it out without making a mess."

Myles Rademan says, "We're talking about our home here. I see this as a community struggle, not as an environmental fight." Which is why he led a trip to see Orville Schell in Bolinas, Calif., in summer 1978 with Messrs. Mitchell, Kuziak, Leinsdorf and Terry Hamlin, the AMAX public-relations man. Mr. Schell's book "The Town That Fought to Save Itself" chronicles a small community's battles with developers for control of its charming Pacific Coast town; and it's a kind of bible in Crested Butte. Myles Rademan wanted the group to meet the author, who commiserated with them. "Here was another community refusing to accept every indignity thrust upon it," says Mr. Rademan.

However, as a town planner, Mr. Rademan believes cooperation with AMAX is the town's best defense. "We're in a position of having to optimize something we don't really want," he explains. "From a professional standpoint, it's exciting. This project could be a model for how you industrialize sensitive areas, as energy and mining companies seek to take over the West. We can set a new level of corporate responsibility."

Corporate responsibility notwithstanding, Stan Dempsey wants to build his mine. But he's not sure it's economically feasible for AMAX to do the project on a scale any smaller than originally proposed. After two years of studies and talk, the big company is beginning to dig in its heels as it attempts to meet an already delayed timetable for construction (work is scheduled to begin in 1981).

And the Crested Butte Town Council won't listen any longer to AMAX's plans for a mine and mill the size the company wants. They are standing fast, as they made clear in their resolution calling for "a stop to the proposed activity on the Mount Emmons project . . . until a real benefit to Gunnison County and the nation can be shown." When Mayor Mitchell gaveled the unanimous vote, the first time town officials went on record against the mine, 250 people attending the meeting whistled and cheered, as television cameras recorded the event.

The gentlemen's war is heating up. On Washington's Birthday, 1979, 30 members of the High Country Citizens Alliance ski-toured over the range to Aspen and held a street protest against AMAX. In early March, two commissioners from adjoining Pitkin County led a delegation from Aspen and cross-country skied back over the high-mountain route to Crested Butte, where they held a rally in opposition to the mine. It's not clear now what is going to happen to Gunnison County and the town of Crested Butte, but the people on the front lines in this fight to determine the future of some of our most beautiful Western lands know what they don't want. The lines are drawn in a battle that will be repeated in dozens of Crested Buttes throughout the country in the 1980's.

Love Canal, U.S.A.

by Michael H. Brown

In the years since Rachel Carson's "Silent Spring," a great national concern has arisen over air and water pollution. It now appears that pollution seeping into the earth itself has gone largely unnoticed and in some cases may be far more dangerous as a direct cause of cancer and other severe human illnesses. "Toxic chemical waste," says John E. Moss, who was chairman of the House Subcommittee on Oversight and Investigations before his retirement this month, "may be the sleeping giant of the decade." Not until the nightmare of the Love Canal unfolded in Niagara County, N.Y., did Americans become aware of the vast dangers of ground pollution. But the problem since then seems only to be worsening.

Each year, several hundred new chemical compounds are added to the 70,000 that already exist in America, and the wastes from their production—nearly 92 billion pounds a year—are often placed in makeshift underground storage sites. Federal officials now suspect that more than 800 such sites have the potential of becoming as dangerous as those at the Love Canal and some are probably already severely hazardous to unsus-

pecting neighbors. The problem is how to find them and how to pay the enormous costs of cleaning them up before more tragedy results. So far, Federal, state and local governments have been, for the most part, reluctant to face the issue.

Sometime in the 1940's—no one knows or wants to remember just when—the Hooker Chemical Company, which is now a subsidiary of Occidental Petroleum, found an abandoned canal near Niagara Falls, and began dumping countless hundreds of 55-gallon drums there. In 1953, the canal was filled in and sold to the city for an elementary school and playground (the purchase price was a token $1), and modest single-family dwellings were built nearby. There were signs of trouble now and then—occasional collapses of earth where drums had rotted through, and skin rashes in children or dogs that romped on the field—but they were given little thought until the spring of 1978. By then, many of the homes were deteriorating rapidly and were found to be infiltrated by highly toxic chemicals that had percolated into the basements. The New York State Health Department investigated and discovered

*This infra-red photo shows the ground pollution in the schoolyard
and backyards in the Love Canal area.*

startling health problems: birth defects, miscarriages, epilepsy, liver abnormalities, sores, rectal bleeding, headaches—not to mention undiscovered but possible latent illnesses. In August, President Carter declared a Federal emergency. With that, the state began evacuating residents from the neighborhood along the Love Canal, as it is named after the unsuccessful entrepreneur, William Love, who built it in 1894. Two hundred homes were boarded up, the school was closed and the nation got a glimpse of what Senator Daniel Patrick Moynihan called "a peculiarly primitive poisoning of the atmosphere by a firm."

But it was clearly not so peculiar. Since then,

new dumping grounds have been reported in several precarious places. Under a ball field near another elementary school in Niagara Falls health officials have found a landfill containing many of the same compounds; it was discovered because the ball field swelled and contracted like a bowl of gelatin when heavy equipment moved across it. Officials have discovered, too, that Hooker disposed of nearly four times the amount of chemicals present in the Love Canal several hundred feet west of the city's municipal water-treatment facility, and residues have been tracked inside water-intake pipelines. Across town, near Niagara University, a 16-acre Hooker landfill containing such killers as Mirex, C-56 and lindane—essentially chemicals that were used in the manufacture of pest killers and plastics—has been found to be fouling a neighboring stream, Bloody Run Creek, which flows past drinking-water wells. About 80,000 tons of toxic waste are said to have been dumped there over the years.

Still worse, as the company recently acknowledged, Hooker buried up to 3,700 *tons* of trichlorophenol waste, which contains one of the world's most deadly chemicals, dioxin, at various sites in Niagara County between 1947 and 1972. Investigators immediately sought to determine whether dioxin had seeped out and, indeed, the substance was identified in small quantities within leachate taken from the periphery of the Love Canal, an indication that it may have begun to migrate. There are now believed to be an estimated 141 pounds of dioxin in the canal site—and as much as 2,000 pounds buried elsewhere in the county. The Love Canal is above the city's public water-supply intake on the Niagara River but a quarter of a mile away; the other sites are closer—in one case within 300 feet—but downstream of the intake. However, the Niagara flows into Lake Ontario, which Syracuse, Rochester, Toronto and several other communities make use of for water supply. Although health officials regard the dioxin discovery as alarming, they do not yet consider it a direct health threat because it is not known to have come into contact with humans or to have leached into water supplies. Academic chemists point out, however, that as little as *three ounces* of dioxin are enough to kill more than a million people. It was dioxin, 2 to 11 pounds of it, which was dispersed in Seveso, Italy, after an explosion at a trichlorophenol plant: Dead animals littered the streets, hundreds of people were treated for severe skin lesions and 1,000 acres had to be evacuated.

☐

Two weeks ago, New York State health officials began to examine and conduct studies of residents and workers in the Niagara University area because of the dioxin concentrations. One local physician there expresses concern over an apparently high rate of respiratory ailments, and union officials say that workers in industries alongside the landfill are suffering from em-

physema, cancer and skin rashes. Cats have lost fur and teeth after playing near Bloody Run; some young goats have died after grazing on its banks, and the creek is devoid of all aquatic life.

So far, there are at least 15 dumps in Niagara County alone that have been discovered to contain toxic chemicals. But no one in the county, or anywhere else in the country, is sure exactly where underground dumpsites are. Of the thousands of covered pits suspected of containing toxic wastes in the United States, the U.S. Environmental Protection Agency says it is a fair estimate that as many as 838 are, or could become, serious health hazards. But the machinery to carry out the kind of monitoring and inspecting now being done in Niagara County does not generally exist elsewhere. And the E.P.A., internal

A young Love Canal resident sits dejectedly by a fenced off area considered to be dangerous by the New York State Health Department.

112

memorandums reveal, has not been eager to set it up because of the extraordinary expense and political problems that would inevitably present themselves. In fact, one regional official was reprimanded for trying to get the type of action that must be taken to guard against another Love Canal.

In at least one known case there are symptoms disturbingly similar. Just 400 feet from a residential area in Elkton, Md., is a disposal area that, according to E.P.A. files, was used both by the Galaxy Chemical Company and by a suspected, unidentified midnight hauler. Residents have complained of sore throats, respiratory problems and headaches, all reminiscent of the early days of trouble at the Love Canal. One local doctor contends that the cancer death rate in the area is 30 times greater than elsewhere in the county, though his report is the subject of much controversy. So far, no evidence of direct human contact with leachate is known to have occurred there, as it did at Niagara Falls; nor have residents demanded evacuation.

In Rehoboth, Mass., 1,000 cubic yards of resins left over from a solvent redistilling process were ordered removed from a dumpsite that the owner had placed within 10 feet of his own house. In Lowell, Mass., some 15,000 drums and 43 tanks of assorted toxic wastes are at present being removed from a site within 200 yards of homes, and chemicals leaking from the drums are appearng in sewers and a nearby river.

Authorities in Michigan claim that the beleaguered Hooker Company has dumped C-56 into sandy soil, contaminating public wells, which have been closed off, and polluting White Lake, near Montague; the state is trying to force the company into a $200 million cleanup. Hooker has also been involved in lawsuits filed by maimed workers in Hopewell, Va., who became sterile and lost their memories after exposure to Kepone, a pesticide that Allied Chemical and Hooker jointly made and packaged.

The U.S. Comptroller General, at the request of Congress, has mapped out stretches through much of the East, Texas and Louisiana and parts of Oregon, Washington and California as regions with the greatest potential for trouble. "Texans are only now becoming concerned about solid-waste disposal," says Doris Ebner, environmental manager for the Houston-Galveston Area Council. "What will happen is that there will be some disaster to make a flash." But serious ground investigations are still not given a top priority.

□

The tendency not to connect health problems with ground pollution has certainly been widespread. In the past, ground pollution was not a major concern in Niagara Falls, either. Because that city is relatively small and has a cheap source of hydroelectricity for chemical firms, most of its people have lived their whole lives on top of or near the hidden strains and goo of industrial pollution. Children near the chemical dumpsites often played with phosphorescent rocks, which would explode brilliantly when they were thrown against concrete. Dirt on the old canal had turned white, yellow, red, blue and black; rocks were orangish; and cesspools of caustic sludge gushed from several locations. These manifestations were viewed more as a matter of esthetics than as a health problem.

But indiscriminate dumping, dumping whatever wherever, has been a national way of life. Though American manufacturers of plastics, pesticides, herbicides and other products that produce huge amounts of toxic wastes are beginning to deposit them in centralized landfill sites—which may insure a closer inspection—the common practice has been to dispose of residues and forget about them. This has been true of private individuals as well, from independent haulers to local farmers.

Farmlands, because they make for nicely isolated dumping grounds, have posed special problems. In 1974, a 100-square-mile pastureland around Darrow and Geismar, La., was found to be contaminated with hexachlorobenzene (HCB), which was produced by the volatilization of wastes dumped into pits. HCB, a byproduct of the manufacture of carbon tetrachloride and perchloroethylene, causes liver deterioration, convulsions and death. During a routine sampling of beef fat by the U.S. Department of Agriculture as part of the Meat and Poultry Inspection Program, 1.5 parts per million of HCB was tracked in the meat of a steer belonging to W. I. Duplessis of Darrow. Further samplings showed that cattle were carrying the same toxin. Soil and vegetation were likewise tainted. The dumps were covered with plastic and dirt, and 30,000 cattle were ordered destroyed. The cattle were fed special diets instead of being slaughtered, however, and moved away from the area; their levels receded to an "acceptable" point, and only 27 were deemed unmarketable and killed. No one can be sure how many cattle, grazing near dumpsites elsewhere, have made it to the dinner table undetected.

Several years ago in Perham, Minn., 11 persons suffered arsenic poisoning from leaching grasshopper bait. Those struck with contamination worked for a building contractor who drilled a well 20 feet from where bait had been buried by a farmer 30 years before. Severe neuropathy cost

one of the employees the use of his legs for six months.

Much of past dumping has been plainly illegal. New Jersey, one of the most industrialized states and one whose cancer rate has been found to be substantially higher than the national average, has been a favorite spot for midnight haulers, or "scavengers," paid to cart off wastes and unload them in swamps, sewers, pits or abandoned wells to avoid paying for disposal at approved sites. In Coventry, R.I., officials found an illegal and highly toxic dump on a pig farm owned by a convicted gambler. It contained one suspected cancer-causing agent, carbon tetrachloride, and another compound that will ignite at 80 degrees Fahrenheit.

Chemical wastes are not always disposed of properly, causing health hazards for future generations.

New York State Department of Health researchers testing the soil in the Love Canal area.

More blatant violators have been known simply to loosen tank-truck valves and get rid of contaminants along roadways in the dark of night. The owner of a New York company that reprocesses electrical transformers is currently on trial on charges of deliberately spilling out polychlorinated biphenyls (PCB's) from his truck onto 270 miles of a highway in North Carolina, and 700 residents of the Warrenton, N.C., area recently protested a state plan to create a new dump there for some 40,000 cubic yards of the tainted soil.

In other instances, dangerous conditions have been brought about more innocently. In Missouri, dioxin was discharged into waste oil and the oil was later sprayed on three race tracks and a farm road to control dust. Some 63 appaloosa and quarter horses, 6 dogs, 12 cats and a large number

of birds died as a result. A child who frequently played on the dirt road was rushed to the hospital with severe bladder pains and urinary bleeding.

Ground pollution's greatest threat is to the national drinking supply. More than 100 million Americans depend upon ground water as the major source of life's most vital fluid. Springs and wells, as opposed to rivers and lakes fed by running streams, are the main drinking reservoirs in 32 states. Florida's population, for example, is 91 percent dependent on ground water. Pouring tons of chemicals into the earth can be comparable, in an indirect way, to disposing of poisonous wastes upstream from a municipal river intake.

Chemical landfills never lie dormant. When water penetrates buried wastes, it removes soluble components, producing a grossly polluted liquid leachate that extends out from the dump. Therein resides the danger. Leaching can continue, at any given site, for more than 100 years, picking up dangerous, stable materials and spreading them around a surprisingly large area. E.P.A.'s Office of Solid Waste has guessed that the average landfill site, about 17 acres in size, produces 4.6 million gallons of leachate a year if there are 10 inches of rainfall. In the spring of 1978, the Comptroller General reported that a *billion* gallons of ground water had been polluted near an Islip, L.I., landfill. A contaminated aquifer that was a mile long and 1,300 feet wide spoiled some drinking-water wells, which had to be sealed off and the homes connected to another source. In humid regions, where rainfall exceeds evaporation, the problem is most acute: The more water in the ground, the more leaching occurs.

□

Several years ago the Union Carbide Corporation contracted with an independent hauler to remove an unknown number of drums from its Bound Brook, N.J., facility. Inside were wash solvents and residues from organic chemical and plastics manufacture. Instead of going to the Dover Township landfill, much of the waste was dumped on a former chicken farm in the Pleasant Plains section of Dover. Mr. and Mrs. Samuel Reich had leased the land to Nicholas Fernicola on the assumption, according to case files, that he was in the drum-salvaging business. When the Reichs smelled pungent odors emanating from the property, they investigated the land and found thousands of containers, both buried and strewn about the surface. Additional drums were discovered in a wooded area near the Winding River, four miles away. The drums were hauled away under court order, but the damage had been done. Sufficient quantities of chemicals had already entered the environment, and early in 1974 residents of the area began tasting and smelling strange things in their water. Dover's Board of Health, in emergency action, passed an ordinance forbidding the use of 148 wells and ordering that they be permanently sealed. Although there were no documented cases of illness as a result, it is difficult to determine how many residents had consumed potentially harmful substances before the odors were noted. Equally difficult is determining where and how far the leachate traveled.

The Government itself has been the cause of serious ground-water contamination. Sloppy storage at the Rocky Mountain Arsenal, formerly an Army production center for chemical warfare agents, led to the contamination of 30 square miles of shallow aquifer near Denver and, in turn, to the abandonment of 64 wells used for drinking water and irrigation. Waterfowl in the area died, and poisoned soil turned sugar beets and pasture grasses a sickly yellow. An estimated $78 million will be needed to complete the proposed cleanup, but there is no way of recovering the chemicals that have already escaped. One irrigation well that shows traces of contamination is only a mile south of the city of Brighton's public well field. The arsenal dug an injection well 12,045 feet deep for immediate disposal, but such facilities do little to insure against long-range migration; as it turned out, the well caused earth tremors and had to be closed.

□

In 1976, President Ford signed into law the Resource Conservation and Recovery Act. It may become an important piece of legislation, if the E.P.A. decides to implement it. This new law provides for a hazardous-waste regulatory program, control of open dumping, an inventory of disposal sites, and grants and programs for communities to set up solid-waste management systems. The passage of that law was provoked by the fact that toxic-waste disposal not only has gone unwatched, but is indeed increasing at an alarming rate. The chief reason for the increase is, paradoxically, the imposition of air and water pollution regulations that have stepped up the practice of burying materials in the ground. Issuance of new disposal regulations was supposed to have been made within 18 months of the President's signature, but today the E.P.A. is predicting that they will not be ready before 1980.

Spurred by the Love Canal crisis, Representative Moss's House subcommittee met last fall to determine what was happening to the law. It was a discouraging hearing. Hugh B. Kaufman, an E.P.A. official assigned to look for landfill problems, told the Congressmen that the agency's policy has been to avoid finding such situations. "There were no guidelines in this memorandum [on landfills] for the regional office to alert the public to the potential dangers," Kaufman testified. "In fact, the memo further instructed the regions not to find new problem sites because they might be required to provide this information to Congress and the public." On July 16, according to Mr. Kaufman, Steffen Plehn, head of E.P.A.'s Office of Solid Waste, told him to stop looking for imminent hazards. Mr. Plehn admitted that Mr. Kaufman's statement was essentially true, but the reason, he said, was that jurisdiction for such matters was being defined under the agency's enforcement division while his unit was culling a "data base." The problem, according to Mr. Plehn, was bureaucratic.

As long ago as April 20, 1978, and more than three months before officials recognized the Love Canal as an emergency, Mr. Kaufman wrote

Special precautions must be taken for disposing of hazardous materials such as pesticide containers to prevent future Love Canal-type contamination.

John P. Lehman, E.P.A.'s Hazardous Waste Management Division director, and said it was "imperative" that dumpsites across the country be cleaned up immediately. "We are receiving reports that, for the most part, the state of hazardous-waste management in the U.S. is as bad or worse than it was when Congress passed [the Resource Conservation and Recovery Act]," Mr. Kaufman wrote, "I recommend that we shift our policy emphasis and not close our eyes to the fact that hazardous-waste facilities located in many states are presenting hazards to the public." Neither E.P.A. officials nor regional offices paid much attention to that advice. When Mr. Lehman warned E.P.A.'s regional office for Ohio that a chemical facility in Akron might be an "imminent hazard" (it appeared to be leaking chemicals into drinking wells), the office sent back a pointed note reprimanding him for "loosely" using the term "imminent hazard" and stating that the region's Air and Hazardous Material Division did "not intend to send any person from this office out to inspect the facility at this time." At about the same time, Mr. Plehn wrote Mr. Lehman a memorandum suggesting that he "put a hold on all imminent-hazard efforts."

But the agency cannot be held as the sole culprit. Its large volume of responsibilities—from car emissions to microwaves—is an awesome task. And it often gets little help from state, county and city agencies. The Niagara County health department and the city government did not consider the Love Canal situation an emergency, for example, and, in fact, played down the problem, and the New York State Department of Environmental Conservation did little in the way of investigation. Not until the state Department of Health stepped in was the matter regarded as urgent.

Much of the randomness with which chemical companies have chosen their dumping grounds over the years will no doubt continue until the Resource Conservation and Recovery Act is implemented. Even then the problem will not go away. There is simply no such thing as a totally secure, self-contained landfill, a fact even those in the business admit. "There is no proof a landfill, 100 years from now, won't leach," says Paul Chenard, president of SCA Chemical Waste Services Inc. He says disposal methods have been improved. Pits can be lined with a special plastic. Waste-disposal firms can excavate on clay-based soil, compact the ground, install standpipes to pump out leachate, and slope the final cover to minimize rain infiltration. But the state of the art is new and no one issues guarantees. Many environmentalists feel that only when there is "cradle-to-death" legislation demanding that wastes be rendered innocuous before disposal will the problem be under control, and there are no signs of that happening in the near future.

□

An E.P.A. memorandum has listed more than 32,254 storage, treatment and disposal sites, both on and off industrial premises, as existing in the nation. In an earlier breakdown, California ranked first, with 2,985; Pennsylvania, New York, Ohio and Texas were not far behind. Those statistics, officials emphasize, refer only to *known* sites. And even at the known sites the quality of the treatment is questionable. One estimate is that less than 7 percent of the 92 billion pounds of chemical waste generated each year receives proper disposal. After working in Niagara Falls for several months, Dr. David Axelrod, New York State Health Commissioner, says the overall problems of improper disposal and treatment "are incredibly immense." The Hooker Company—which contends that it did not know the possible dangers and was simply disposing of wastes as everyone else did—is already faced with claims against it in excess of $2 billion, and citizens' demands upon the state are only just beginning. New discoveries of dioxin are prompting new demonstrations, new arrests of demonstrators and new requests for evacuation and relocation. Patricia Pino, whose home in Niagara Falls is now unmarketable, was one of those arrested. "We request a reprieve from death row," she telegraphed Gov. Hugh Carey. "We are innocent of any crime." Her two children have liver abnormalities, and she has learned that she herself has cancer.

The Countless Mysteries of Peatland

by Jane E. Brody

Like the meadow in Andrew Wyeth's painting "Christina's World," the sprawling field of tall grass shimmering in the gentle breeze seemed a fine place to have a picnic—until you stepped into it and sank to your ankles in cold water. For it wasn't a meadow at all, and the rippling vegetation wasn't grass. Rather, it was a plant called sedge, one of hundreds of species that live with perennially wet feet in a soggy peatland that covers 300 square miles of northern Minnesota.

Because of an ever-tightening energy supply, the many layers of peat in Big Bog, as it is called on some maps, are attracting interest as a possible source of natural gas. But to Dr. Eville Gorham, an ecologist at the University of Minnesota who frequently visits Big Bog with a dozen students, the waterlogged terrain is a source of countless mysteries just waiting to be unraveled.

"Somehow, I find it hard to look over this and see nothing but B.T.U.'s," Dr. Gorham said as he gestured across the expanse of sedge, his deep blue eyes reflecting the tranquility of the waving strands of green that were tinged with silver in the late summer sun. Dr. Gorham is one of the very few modern scientists who study the ecology of peat bogs, which are commonly called swamps. They are not wet enough to boat through or dry enough to walk through without getting wet. Dr. Gorham's students say that peatland ecology is a very primitive science for which there is no text. "We read scientific papers and long monographs, take a lot of field trips and do our own studies," said one young woman from Bemidji, Minn., who said that she was taking Dr. Gorham's course because, although she grew up in the midst of peatlands, she knew nothing about them.

The sedge "fen," as the Europeans call this type of peatland, is fed by waters that percolate through mineral soils and by nutrient-rich runoff. But, according to Dr. Gorham, a true bog receives its nutrients almost solely from the atmosphere—from rain, dust, industrial pollution and sea spray—and supports a wide range of vegetation, from sphagnum (peat moss) and lichens to tiny birch and tall black spruce.

A swamp—more a popular than a scientific term—is a forested wetland that can be either a bog or a fen, usually with water standing or flow-

Dr. Eville Gorham, an ecologist from the University of Minnesota, trekking through a sedge fen in the Big Bog.

ing through or over it. A marsh, on the other hand, has few or no woody plants but does have grasses and reeds in its silty soil.

A DISTINCTIVE ECOLOGY

The unschooled observer may not see anything special about peat bogs. They are not as eerie as the great swamps, nor do they teem with apparent wildlife. Yet to a scientist like Dr. Gorham, their distinctive ecology makes them a fertile area for deciphering the delicately balanced relationships between plants, animals and their waterlogged environment. They also possess some rare properties that have proved historically as well as biologically fascinating.

In May 1950, a professor named P.V. Glob was called to examine a very well-preserved body that

had just been found in Tollund Fen, a Danish peat bog. The workers who unearthed the body from eight feet of peat thought they had stumbled upon a recent murder victim.

But Professor Glob wrote in his book, "The Bog People," that the body was that of a man who lived 2,000 years ago, a relic of the Iron Age preserved by the acidic, antiseptic, oxygen-depleted waters of the bog. More than 700 such ancient bodies have been dug out of the bogs of northern Europe, all remarkably well-preserved.

Most peatlands develop in flat areas where the climate is cool and damp and the water supply is subject to damming, often by beavers. Whether any bodies would be found in northern Minnesota is anybody's guess. But what can be found are layers of sphagnum moss so well preserved that moss two or three meters down can be identified as to species, Dr. Gorham said.

A SPONGE-LIKE QUALITY

Sphagnum is the signature plant of a bog. Its large water-holding cells give it a sponge-like quality that allows it to absorb 200 times its own weight in water. Although most commonly used to condition the soil in gardens and house plants, sphagnum has also served as litter for house pets, baby diapers and, in World War I, as wound dressings, the scientist said.

When compacted into peat and dried, sphagnum has twice the heating capacity of an equivalent amount of wood. Peat is formed in cool, waterlogged places where the lack of oxygen inhibits bacterial decay. Sphagnum further resists decay because it is highly acidic and contains nitrogen in a form not usable by bacteria.

Beyond the reeds and cattails along the roadside ditch, the sedge fen seemed at first to be a monotonous stretch of grasslike greenery. But, as Dr. Gorham slushed through the sedge, he exposed the incredible variety of life hidden in the bog.

Parting the sedge, he unveiled a clump of bladderworts, small flowering plants with tiny pouches that sit in the water and catch and digest microscopic animals that happen past. Then he sighted an insectivorous plant, the sundew, a delicate rare species with sticky droplets that glisten in the sun.

Sphagnum moss can absorb 200 times its own weight in water.

THE COUNTLESS MYSTERIES OF PEATLAND

CUSHIONS OF SPHAGNUM

Scattered here and there were patches of mosses, marsh fern, irises, low willow and buckbean, which has airholes that carry oxygen to the water-logged roots to keep them alive. Deeper into the sedge fen, cushions of sphagnum began to appear, creating spongy hummocks raised above the water surface where plants that can't live in water can get a "roothold." The highly poisonous water hemlock, leatherleaf, cranberry, sweet gale, bog rosemary and bog birch were among the many species of plants living on the sphagnum hummocks. Finding one large hummock honeycombed with the runs of small mammals, Dr. Gorham said, "This is the animals' island in the sea." He said that large populations of weasels, moles, mice and rabbits lived in the wetter areas (a marsh hawk hovering overhead was looking for one for lunch), and that moose and deer lived in the forested fens. But, he added, no systematic study has been made of animal life in peatlands.

Many of the hummocks also housed pitcher plants, another rare species, with leaves like rams' horns that catch and hold rainwater. Insects that crawl into the pitchers are trapped by downward-pointing hairs and then digested by a plant enzyme. A species of mosquito has evolved that is resistant to the enzyme and uses the water in the pitchers as breeding pools.

SMALL TREES FOUND

Still farther into the bog, larger and drier hummocks were populated with small, shallow-rooted trees—tamaracks and black spruce, with labrador tea, lichens and lingonberries at their feet. Thoroughbred Big Bog, teardrop-shaped "tree islands," their long axes parallel to the direction of waterflow, dot the flat landscape.

Now it was time for the real work. Dr. Gorham and his students sank metal corers deep into the hummocks to study their evolution. They circumscribed areas in which to count the plant species. They took the temperature of the soil and light levels at the surface, measured tiny changes in the soil and water levels and collected water samples to analyze their chemical contents.

In previous studies, carbon-dating of the bottom layer of peat 85 inches down showed it to be 4,360 years old, indicating that about two inches of peat accumulated each century, a much slower rate than the present peat buildup.

The Energy-Efficient Society

by Anthony J. Parisi

Welcome to the efficient society.

Almost unnoticed, Americans have begun to save energy by wasting less. In the process, they have almost halted the growth in oil imports, lessened the drain on the dollar, deflected some inflationary forces, avoided a lot of pollution and eased the tension over such issues as nuclear power and strip mining. The nation remains in a difficult energy bind—there is absolutely no question about that—but these unheralded gains suggest that the country is, in fact, stumbling on a way out of that bind, one that is far less tortured and painful than many thoughtful people had feared. It is called energy efficiency.

"We are already on the road to solving our energy problem," says Vince Taylor, a physicist and economist who is a consultant to the Union of Concerned Scientists, an enviromentalist group. "The path is really bright. The opportunities are enormous. Ten years from now we will look back and be amazed at how much energy we've been able to get along without—and without major changes in our life style."

To many, Taylor's words may sound like opti-mism run amok. But the facts he marshals are impressive. In a paper entitled "The Easy Path Energy Plan," which is fast making the rounds of the energy community, Taylor opposes the conventional view as held both in the United States and abroad—that America is the land of the free and the home of the gluttonous. He shows that in 1978 America consumed just 5 percent more energy than it did in 1973, even though the economy expanded more than 12 percent during those five years. By contrast, over the prior five-year period it took a 22 percent increase in energy consumption to fuel an economic expansion of 17 percent. If Americans had been as energy inefficient in 1978 as they were in 1973, the nation would have had to import half again as much oil as it did.

"The reason our energy problems appear so intractable to most people, even to supposed experts, is that attention has been focused entirely on the possibilities for expanding supplies," Taylor concludes. "Yet the more closely people have looked at these supposed solutions, the clearer it has become that this route will lead to no immediate answer." Meanwhile, conservation has been largely written off in official circles as back-to-the-woods romanticism. Or, worse, those wedded to the notion of infinitely expanding supplies equate conservation with sacrifice and burden. Yet sacrifice is something the efficient society tries to avoid. The need and desire to overcome human hardship were, after all, motivations for creating social organization in the first place.

□

There is now evidence to suggest that many Americans are realizing that they have a choice, in dealing with the energy crisis, between suffering and energy efficiency, and they are choosing efficiency. Public posturing aside, they largely ignore the 55-mile-per-hour speed limit and balk when told to bear 65 degrees in winter and 78 degrees in summer. Instead, they are driving smaller cars, insulating their homes, streamlining their industrial processes and designing their buildings to ward off the sun in summer and embrace it in winter.

America is beginning to capture energy once wasted, and people are discovering in the process that there are rewards beyond even the money this saves, much like the little thrills one gets from enjoying leftovers or finding a bargain. "There are a lot of subtle, intangible benefits," says Amory Lovins, of the Friends of the Earth, whose scholarly, almost poetic, papers have made him a kind of high lama of the energy-efficiency sect. "It makes you more flexible," he says. "It's good for the spirit."

It's also effective. Heating specialists have found that reducing the temperature of a poorly insulated home from 68 to 65 degrees will trim fuel consumption by only about 15 percent. Adding more insulation in the attic and other accessible places can cut consumption in half. With a comparable new home, thoroughly insulated, designed with such passive solar features as greenhouses to trap the sun's energy in winter and southerly overhangs to block it in summer, and perhaps fitted with a modern wood-burning stove instead of an anachronistic fireplace that literally sucks heat out of the room, a family can easily halve its fuel consumption yet again.

Driving slowly in a car that gets 15 miles a gallon helps a little; driving a car that gets 27.5

miles a gallon, the level mandated for 1985 models, helps a lot. Based only on the mileage standards already on the books—standards that many believe are too loose—the oil industry expects the demand for gasoline, which accounts for 40 percent of all the oil consumed in this country to start dropping within a year. Some are already saying that we may never again use as much as we did in 1978.

New skyscrapers can provide all the comforts of buildings designed in the 1960's but with half the energy. Among other innovations, they recycle waste heat from computers, lights and even the occupants. Those precious B.T.U.'s used to be thrown away—even as the boilers below were devouring still more fuel to replace them.

In the industrial sector, where 40 percent of the nation's energy is consumed, the spectacular gains in energy efficiency have already crushed old canards. Despite repeated warnings from business quarters that zero energy growth might mean economic disaster, the nation's mills and factories actually used less energy last year than they did in 1973. Dow Chemical, for example, realized in the 60's that rapidly rising fuel costs were putting its Michigan plants at a competitive disadvantage to petrochemical complexes on the Gulf Coast, where energy was still considerably cheaper. In 1967, Dow, which alone devours .5 percent of all the energy consumed in the United States, launched an aggressive fuel-conservation program. Within a decade, it cut the amount of fuel needed to make a pound of product by 40 percent, principally by plugging steam leaks and keeping them plugged.

□

Some experts believe that industry will be able to get by on less energy in 10 years than it needs now, especially if the Government grants greater tax credits on investments that save energy. But even without such encouragement, advances are coming. The Union Carbide Corporation, for example, has started making polyethylene, a petrochemical mainstay, with a new process that uses only a fourth as much energy as the old.

All these gains are quietly adding up to a staggering sum. "Since 1973," says Lee Schipper, a staff scientist at Lawrence Berkeley Laboratory, "we have gotten more energy from efficiency im-

These aluminum beverage cans will be recycled at a saving of ninety-five percent of the energy needed to make aluminum from bauxite.

Solar water heaters are a standard feature of this housing development in California. Well insulated homes increase the viability of such systems.

provements than from any form of supply—including oil imports."

Ironically, it was exactly these kinds of improvements that President Carter anticipated when he proposed his first, conservation-oriented energy plan three years ago. Congress, sensing public dissatisfaction, chopped it to bits. Now the President is back with a supply-oriented plan centered not on efficiency but on synthetic fuels. The benefits of energy efficiency are even more appealing today, however, than they were then.

These advantages tend to cluster in two areas: economics and the environment. The environmental appeal is obvious—energy recovered through efficiency improvements leaves few if any traces—but the economics are less apparent. The A.F.L.-C.I.O.'s George Meany and the leaders of the National Association for the Advancement of

Colored People are among those who have opposed an all-out program for conservation and energy efficiency, and the essential thread of their objections is that it would somehow result in deprivation, especially for lower-income groups. Thus, some workers and minorities have formed a loose coalition with some industrialists who like to crow that the United States produced, not conserved, its way to greatness.

But this conventional faith in production and ever-expanding energy consumption as the means to achieving all social ends went awry as fuel costs started soaring in the early 70's. When that happened, it became cheaper to conserve than to produce. The challenge posed by the energy crisis is not how to get out of the bind for free—that's impossible—but how to get out the least costly way. As Amory Lovins puts it, "Conservation will not be cheap—just cheaper than not conserving." The key, he says, is capital: "These days, it takes less capital to save energy than to produce it. Society has only so much capital, and what you want to do is invest it in the most productive way possible. If we fail to do that, money will be drawn unnecessarily from other uses—including a lot of social programs that might have helped poor people." As far as jobs are concerned, the nature of some industries that are tied to energy conservation is such that they themselves become substantial employers. "A dollar spent on insulation creates more jobs than a dollar spent on shale oil," Lovins points out.

☐

Viewed in this perspective, efficiency improvements begin to look like a steal compared to such big technology schemes as converting coal into gas and oil, the alternative that President Carter has advocated. By the latest estimates, commercial versions of these synthetic fuel plants would produce oil costing around $40 a barrel, a figure that has tripled in the last five years. By contrast, researchers at the University of California, who are developing housing efficiency standards for the Department of Energy, conclude that a national program to insulate the nation's existing homes would recover energy at a cost of between $10 and $20 a barrel. There would be no environmental disruption, and the savings would start pouring in right away, not 10 years from now. "For $1,500 to $2,000 dollars," says Arthur Rosenfeld, a University of California physicist, "we could retrofit the typical American home and cut energy consumption in half—in half."

Savings of this magnitude may be possible in the economy as a whole. Sweden and West Germany now consume 50 percent less energy for the same amount of heavy industrial production than the United States does. The situations are not precisely comparable but there is no fundamental reason why Americans, given time, cannot become at least 30 percent more efficient.

Roger Sant, director of the Energy Productivity Center of the Mellon Institute in Washington, D.C., has compiled a breakdown of what the country's energy picture might have looked like in 1978 if, magically, all energy services had been delivered, not by the expensive ways we have inherited, but by the cheapest, most efficient means possible. The findings, he says, are "fabulous": Oil consumption would have been 28 percent less; imports would have plunged from more than eight million barrels a day to barely four million; well over a third of all central power stations could have been eliminated, perhaps obviating the need for nuclear power; overall energy costs to the consumer would have dropped 17 percent.

Even before the first oil shock, in late 1973 and early 1974, energy efficiency had become the subject of considerable intellectual scrutiny. The Ford Foundation's controversial 1974 report "A Time to Choose," for example, warned that if the United States continued using energy at the going rate, supplies would have to swell an impossible 150 percent by the turn of the century. "Technical fixes" could hold energy growth to 50 percent, the report declared, with no appreciable impact on economic growth. And if the nation went all out, the study added, America could achieve zero energy growth early in the 21st century.

In business circles, the idea of zero energy growth went over like Copernicus in the house of Ptolemy. Traditionalists charged that the study was unsound and the studiers were incompetent. Yet today most projections for domestic energy consumption 20 years hence—even most industry projections—are closer to the foundation's zero-growth scenario than its technical-fix scenaro. Some have gone beyond: early last year the President's Council on Environmental Quality issued a paper with the happy title "The Good News About Energy." It concluded that total energy use in the United States "need not increase greatly between now and the end of the century, perhaps by no more than 10 to 15 percent."

☐

Last year, none other than the Harvard Business School, no stranger to business interests, endorsed the efficiency-first energy path. In a report entitled "Energy Future," it concludes: "The United States can use 30 or 40 percent less energy than it does, with virtually no penalty for the way Americans live—save that billions of dollars will be spared, save that the environment will be less strained, the air less polluted, the dollar under less pressure, save that the growing and alarming dependence on OPEC oil will be reduced, and

Solar collectors form the roof of this waste treatment plant in Wilton, Maine, and supply heat to sludge digesters, enabling the production of methane gas. Thus a waste product generates an energy resource.

Western society will be less likely to suffer internal and international tension."

As the Harvard study makes clear, there is another, more pragmatic reason for favoring efficiency. Except for the slow progress being made on solar and geothermal forms of energy, all other domestic energy sources either seem to be waning naturally or wallowing in an environmental muddle. And even solar enthusiasts, generally an impatient breed, concede that before society could truly switch to renewable sources, it must trim waste. "We should recognize," says Denis Hayes, director of the federally funded Solar Energy Research Institute, "that conservation is a precondition to a solar transition." Solar becomes a much more attractive option in an energy-tight home, for example.

Despite the sharply higher prices that oil and gas producers have enjoyed in recent years, the nation's proved reserves of oil have declined 28 percent since the record level reached in 1970; gas reserves have dropped 32 percent from their peak in 1967. The declines have come in spite of a near doubling in the number of wells drilled since 1971. Hoping to spur still more exploration (as well as to encourage more conservation), the Government is now phasing out price controls on both these fuels. Few experts think this move will do more than delay the long decline in these once-abundant hydrocarbons.

Coal production is rising, but slowly, and quality has been slipping. On a tonnage basis, Americans consumed only 9 percent more coal in 1978 than they did in 1973, and on an energy-content basis, the increase was less than 7 percent, because the growth has come mainly in low-quality Western coal. The energy from this additional tonnage did not even offset a fifth of what was lost from declining oil and gas production. Dreams of doubling or even tripling coal output by 1985 have long since faded. The supply of coal is there, but the demand is not, at least not on terms that the public finds acceptable. Although the United States has nearly a third of the world's coal, a series of concerns, mostly environmental, have restrained development at virtually every stage. The biggest: resistance to more strip mining; labor, safety and health problems in underground mines; and, most telling of all, air-pollution regulations where the extra coal would be burned. Converting coal into oil or gas, in addition to being very expensive, poses fresh environmental concerns that some think could prove even more troublesome.

Nuclear power, after a slow start, has grown rapidly in recent years and now provides close to 13 percent of all the electricity and nearly 4 percent of all the energy consumed in the United States. Long before the accident at Three Mile Island dropped a big question mark in the atom's path, however, new orders for nuclear power plants in this country succumbed to relentless environmental, safety and even economic pressures; orders have all but halted. Because of these pressures, the Harvard study concluded that nuclear power "could actually undergo an absolute decline within 10 years."

Some of the objections to nuclear power and other new energy projects could be overruled, of course. President Carter presumably had that in mind when he proposed an Energy Mobilization Board. Yet this tactic might mean only more and more confrontation.

□

"Carter described the problems as environmental red tape, but that is a fundamental misconception," Luther Gerlach, professor of anthropology at the University of Minnesota, insists. For 10 years Gerlach has studied what he calls the real "energy wars," the protests that have increasingly erupted around the country over new energy projects. He believes it is foolhardy to assume that these struggles are the handiwork of a small group of dissidents. "People say the Tellico Dam was stopped by the snail darter, but it wasn't," he says. "It was stopped by people who didn't want to lose their land and who seized on the Endangered Species Act as a way to prevent it from happening. We have devised a ritual for doing this

sort of thing, and the ritual isn't the problem. In fact it has helped keep these things from getting out of hand. Take it away," he warns, "and there will only be more direct conflict."

Denis Hayes of the Solar Energy Research Institute believes, as do others, that, in this sense, energy efficiency is not only the cheapest, cleanest and fastest way to go today but that it is also the most harmonious. Greater social harmony emerges as a kind of grand bonus. The alternative, as Amory Lovins has put it, is a highly managed society ruled by a "complex of warfare-welfare-industrial-communications-police bureaucracies with a technocratic ideology."

And yet, technology pervades the efficient economy. The difference is that it is servant, not icon. It may be as simple as a new formula for caulking or as complex as a microprocessor that improves auto mileage by automatically adjusting the carburetor's fuel-air mixture. But it is usually unobtrusive, which is why the sizable strides in efficiency already achieved in the American economy have gone largely unnoticed. America, perhaps more than any other nation, has liked its technology big. A crash program gave us synthetic rubber during World War II, so why can't a crash program give us synthetic fuel now? Economists, in their own way, often concur. When the price of a resource rises, they reason, technology's scope widens, marginal deposits eventually become economical and supplies expand. Why not now?

The answer is that high energy prices are having much more clout on the demand side of the equation than on the supply side. This is what happens when a resource base can no longer outpace consumption and begins to shrink from exhaustion, as domestic oil reserves seem to be doing right now. Technology is responding all right, but not so much in exploiting marginal resources as in finding ways to get by on less. Yankee ingenuity lives.

□

If—and that "if" is still very much warranted —the trend continues, the nation will wean itself from oil as other resources (more coal for a while, then solar) begin to shoulder the burden. In this event, oil prices will eventually taper off at some lofty level and oil will linger on as a highly expensive commodity reserved for only the most critical purposes. Just as whales outlived the importance of whale oil, petroleum deposits will outlast dominant use of petroleum.

A silo fills a 110-car train with coal as it inches through the structure. It takes approximately two hours to load the train that never stops during loading, thereby increasing energy efficiency.

All this might be taken for granted except that generations of Americans have enjoyed nearly a century of expanding energy supplies and shrinking energy prices. Americans have a certain mind-set toward supply, a bias imbedded not only in their thinking but in their customs, laws and institutions. The examples are endless: Federal subsidies for energy-development projects, tax depletion allowances, accelerated depreciation for power plants, write-offs for business fuel expenditures. By conservative count, the pricing subsidies alone may exceed $50 billion a year. Along the way, Americans seem to have lost the kind of energy sensibility that scarce supplies and high prices exact over time. Amory Lovins gives an illustration:

"Thirty years ago, refrigerators used to have motors that were close to 90 percent efficient. Today they're 60 percent or less, because as electricity prices fell, we skimped on the copper windings. We used to put the motor on the top, where it belongs; today it goes underneath, where it must use half its energy just to get rid of its own heat. The trim-line craze hit, so we skimped on insulation—remember when refrigerator walls used to be four inches thick? Besides that, we design them so that when we open the door the cold air falls out. We could make them horizontal, like supermarket freezers, or at least have individual compartments to keep some of the cold air in. Because we let the cold air out so freely, frost builds up quickly, so we put about 700 watts of heaters in most of our refrigerators to get rid of it. A lot of models also have strip heaters around the doors to keep the seals from sticking; a Teflon coating would do. Then to get rid of all this heat, we stick a radiator on the back, and let it heat up the house to give our air-conditioners something to work on. To top it off, we usually install the refrigerator next to another heat-generating appliance such as a stove or dishwasher, so when that goes on the refrigerator tends to go on, too."

□

To be sure, the Government now sets targets for energy efficiency in appliances. It has also tried to offset the subsidies for supplies that are by now tightly woven into the economy by offering tax credits for home insulation, for example. But habits acquired over decades are not easily discarded, and the deck clearly remains stacked against efficiency. "There just isn't any constituency for conservation, despite all its appeal," says Robert Stobaugh, the director of the Harvard Business School study. "That's why the Government has got to get behind it more than it has." He argues that the nation must devise a conservation system to compete directly with oil, gas and electricity, and suggests one way would be to get the utilities to coordinate energy-saving investments in homes, collecting half the money from the Government and half from the homeowner.

□

A few utilities have already moved in this direction. Northern California's Pacific Gas & Electric Company, one of the largest in the country, has for the last year and a half offered its customers loans of up to $500 for ceiling insulation—and the company says customers typically reduce their heating bills by 25 percent. That alone is almost enough to cover the payments on the five-year, 8 percent loans. The rather dramatic result is that with no money up front and only a modest monthly charge, customers can improve their home insulation, and, then, five years later, start pocketing at least a fourth of what they would have been paying for heat.

The company expects 10,000 customers to take advantage of the plan this year. That, it figures, will

free enough energy next year to supply 970 average gas customers and 223 average electricity customers. "If we can put enough of these savings together," R. Michael Mertz, manager of Pacific's energy and conservation services, says, "it will turn out to be a lot cheaper than building a new power plant."

In western Oregon, the Portland General Electric Company has taken the idea one small but psychologically important step further. It will pay contractors to install insulation, caulking, storm doors and whatever else it takes to reduce power consumption in the 90,000 or so electrically heated homes that the utility serves. Customers do not repay the interest-free loans until they sell their homes. Meanwhile, the money is included in the utility's rate base, where it earns the same return as the company's investments in power plants. This ground-breaking arrangement neutralizes management's natural inclination to expand the rate base by building a new plant even when money-saving alternatives abound. "I think this is one of the most exciting things happening," says Roger Sant of the Mellon energy center. "There's this attitude that we ought to keep the utilities out of it, and I think that's unwise."

More such imaginative approaches to improving energy efficiency in the residential sector are needed.

The bicycle remains one of the most efficient forms of transport devised by man, and is being used increasingly by commuters.

THE ENERGY-EFFICIENT SOCIETY

With just a little enticement, industry already seems eager to grow more efficient on its own, because efficiency, in a future of high energy cost, will be the only road to profitability. Savings in the transportation field can also mount up quickly, because the national fleet of motor vehicles lasts six years on the average, meaning that mileage standards can be raised as warranted and can be broadly implemented within that time. Office buildings and homes stand for decades, however. Even if new construction becomes a model of energy thrift, can the country afford to ignore its existing housing stock? Within those structures lie "'proved reserves" of energy that it has barely begun to tap. Tapping them will not be easy—just easier than not tapping them. The costs

will occasionally be steep, the conceptual barriers formidable, the politics exasperating. But the alternative is the inefficient society, which, we are finding, is a concept that has suddenly grown obsolete. "This country was built on inexpensive energy, and we had plenty of it domestically until the early 70's," says David Sternlight, chief economist for The Atlantic Richfield Company. "The situation has changed, and now that it has, we are beginning to change, too."

The heartening thing in all this is that the solution to the energy crisis, if one can hope that these beginnings really will be pursued, seems to be emerging in spite of ourselves. And there is, it turns out, a certain simple elegance to it all.

Cancer Safety

by William E. Burrows

Sometimes a simple question put by a layman can get to the heart of a complex scientific problem more surely than expert interrogation. So it was at the New York Academy of Sciences' International Conference on Public Control of Environmental Health Hazards in 1978. Peter A. Berle, then New York State's Environmental Conservation Commissioner, had just unveiled a plan for dredging heavily contaminated "hot spots" along the upper Hudson River of 340,000 pounds of the PCB's that had been dumped into the waterway by the General Electric Company. A succession of health officials, scientists and corporate representatives rose to question Mr. Berle about the details of removing the CB's (polychlorinated biphenyls—heat-resistant chemicals used for insulating transformers and capacitors), which have been shown to cause birth defects, reproductive failure and cancer in animals, and are strongly suspected of being carcinogenic to man.

When her turn came, Carolyn Cunningham, the head of the Conservation Advisory Council for Rye, N.Y., whose "Tuesday Only" tag indicated her limited status at the meeting, stepped to the microphone and asked a question so elementary that none of the experts in the room would have

dreamed of raising it. Yet the answer to it continues to elude and confound them all: "The acceptable level for PCB's has been set at 5 parts per million. Is that level safe?"

Mr. Berle deferred to Leo Hetling, then his director of pure-waters research, who answered: "That's like asking how many cigarettes can one safely smoke. The Food and Drug Administration considers it safe for eating fish."

Mrs. Cunningham was not satisfied with Mr. Hetling's evasive reply. But his smoking analogy amounted to the plain truth. No one knows exactly how many PCB's the human body can safely assimilate, so setting "acceptable" levels for them, as well as for thousands of other dangerous chemicals, including saccharin, is scientifically absurd.

□

Saccharin, in fact, is currently the object of a bitter controversy over whether it should be banned in the United States, as it is in Canada. Repeated laboratory tests have showed that the artificial sweetener causes cancer in animals and almost without question can do the same in humans. While acknowledging that saccharin can be lethal, a National Academy of Sciences panel recommended to Congress that the additive be restricted but not banned entirely, since it is considered necessary for those who must restrict their sugar intake. There were 12 vigorous dissenters among the 37 panel members, however. Some of them went so far as to contend that there is really no scientific basis for setting human risk levels for carcinogens and that dividing additives into risk categories is, therefore, senseless. Meanwhile, the House Subcommittee on Health and the Environment scheduled hearings on saccharin amid what was one of the fiercest lobbying battles in memory.

□

Levels of acceptability—the doses of chemicals and radiation people may safely absorb—are becoming critically important now that so many cancers are thought to be environmentally caused. And since most cancer-causing chemicals have long latency periods, the real effects of the petrochemical explosion of the 40's, 50's and 60's are only just beginning to be felt. An estimated 400,000 Americans will die of cancer this year and it is thought that 20 percent of all Americans now living—perhaps one out of every five reading this—are going to die of some form of cancer unless something is done about it, and quickly.

Yet none of the public-health officials who set acceptable levels for the thousands of chemicals we eat, drink and breathe can be absolutely sure that those levels adequately protect the public from serious illness or deadly disease. Any dose of a toxin, however low, involves some risk. But trying to calculate the precise degree of risk is incredibly difficult.

To make matters worse, the entire level-setting process takes place in a regulatory morass involving several Federal agencies and a dozen laws (each of which is to some extent vague, inconsistent or contradictory), state and local agencies, the courts, Congress, business interests, environmentalists, "independent" testing laboratories, the National Academy of Sciences, hospitals and universities, a legion of lobbyists, and a growing number of health functionaries on virtually all levels of government. The result is pandemonium.

In the end, economic, not scientific, factors frequently prevail. Often these are incorporated into

law. The Toxic Substances Control Act, for example, stipulates that health risks be balanced against economic factors and other "public impacts." It is the scientists and statisticians, however, on whom we must depend to apply data from cellular and animal experiments in an effort to figure out how much of which poison, individually and in combination, the human body can tolerate. But E. Cuyler Hammond, a vice president of the American Cancer Society and an expert statistician, says that the process, though seemingly sound, is basically "by guess and by God."

Yet industrialists, lawyers, politicians and economists in essence take these numbers, imprecise as they are, and present the public with difficult choices: How much of these poisons are you willing to live with? If you don't swallow a little poison every day—a level we agree is acceptable—you will be thrown out of work and the country will go to the dogs. The strong suggestion is that eating, drinking and breathing a certain amount of toxins is the inevitable price of progress. We are, in effect, being conditioned to accept man-made poisons in our air, food and water as being normal.

The vast majority of people, however, never think of challenging these choices until a Love Canal starts bubbling under their feet or a James River is closed to fishing. Then they react with outrage; not because it happens, but because it happens to them.

□

Lawyers play key roles in level setting because the process is not so much an ongoing clash of opposing scientific truths as one between conflicting social and economic priorities. Regulatory agencies have a real fear of being sued by industry, according to Dr. Sidney Wolfe, a physician who was on the staff of the National Institutes of Health before becoming director of Ralph Nader's Health Research Group in Washington. "In order to keep the agency out of court, they act only when they have the most open-and-shut case," he claims. "When we go into a hearing, we not only see lawyers from industry, but Wall Street analysts, too. There's big money at stake."

What's acceptable and what isn't is most often associated with something called cost-benefit, or risk-benefit, analysis. Deceptively simple in theo-

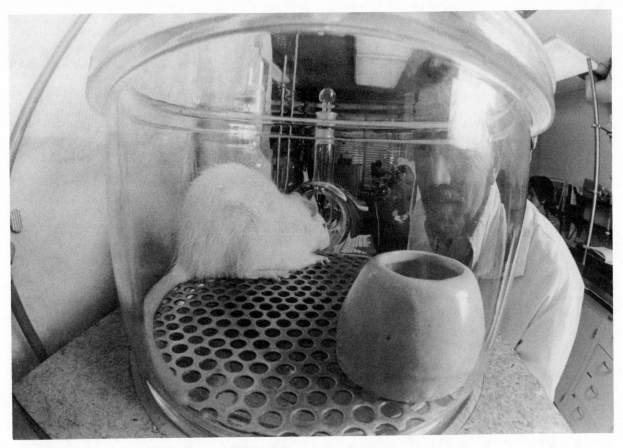

Repeated laboratory tests have shown that Saccharin causes cancer in animals.

ry, it weighs pluses and minuses. The benefits of, say, asbestos are added: fire-resistant homes, schools and offices, plus jobs for asbestos workers, and their combined effect on the economy. The minuses are also totaled: the sharply increasing numbers of people dying from lung cancer, and the cost of caring for the ill and supporting the survivors.

If benefits clearly outweigh costs, it is full speed ahead. If costs far outweigh benefits, the product either never sees the light of day or, if already on the market, is canceled. There are generally no clear-cut imbalances, however, so some level thought to be acceptable is invented.

This raises a fundamental question: To whom, exactly, is the level acceptable? When a newscaster says that "the quality of air today is acceptable," to whom is he or she referring? What is acceptable to the taxi driver who must earn a living with his automobile is not necessarily ac-

ceptable to the asthmatic breathing the fumes—except at the moment when the asthmatic, having a seizure, needs the taxi in order to get to a hospital; then the taxi and its wake of noxious vapors become eminently acceptable to driver and passenger alike.

Although corporate lawyers and economists, as well as many in the public sector, maintain that it is perfectly fair to make production decisions and set acceptable levels after a careful weighing of all likely costs and benefits, others are beginning to challenge the concept on two grounds.

First, cost-benefit tries to equate what is economically, let alone morally, unequatable: kilowatt-hours of electricity produced by a nuclear power plant versus the number of cancer deaths caused by a slow radiation leak at the site where the spent fuel is stored, for example. This kind of equation makes no sense unless we can decide how many kilowatt-hours of electricity a human life is worth.

Second, those who pay the costs and those who derive the benefits are not necessarily the same. Men who dig coal pay a heavy cost in terms of cave-ins, poisoned air and black lung, yet their monetary benefit is not commensurate with those risks.

Before a meaningful scientific basis can be established for setting truly acceptable levels, a deeply important question needs to be answered: Exactly how much of a given substance does it take to cause a human cell to turn cancerous? If no one knows, then every person who believes that there is some magic formula that constitutes a boundary between safety and death is seriously misguided. If there are boundares, they stand not so much between health and illness as between profit and loss.

The vast majority of those in cancer research think that environmentally caused cancer is the result of something called a dose-response relationship. This means that there is most likely some kind of direct relationship between the amount of the cancer-causing chemical a person is exposed to and the length of time it takes to contract the disease. A dose-response chart would have a line going from zero to maximum dose, with the risk accelerating and the time span shortening in proportion to the increasing dose.

The big question, though is this: At precisely what point along that line will someone contract cancer? Given the fact that individuals vary, that there are long latency periods for most cancer-causing substances, and that combinations of chemicals are undoubtedly worse than the simple sum of their elements, it becomes impossible to pick a point on that line and say with certainty that it represents the start of a malignancy. That is what Leo Hetling meant when he asked about the number of cigarettes that can be safely smoked. No cigarette can be safely smoked, but it is impossible to tell which of the thousands upon thousands will be the one that begins the process of wild cellular growth that is cancer.

□

There are currently two principal ways in which scientists try to determine what causes cancer and how much of it is too much: studying groups of humans and testing animals. Neither method is wholly satisfactory. Epidemiology compares groups that are exposed to certain kinds of chemicals and radiation with control groups that are not exposed. It has some serious shortcomings, however. For one thing, it is impossible to isolate people, like laboratory animals, for most or all of their lives, so it is impossible to know about everything they eat, drink and breathe. Further, the cause and effect relationship in epidemiological studies is often weak. Asbestos workers *do* have a higher percentage of lung cancer than does the general public, but that does not necessarily mean the asbestos causes the cancer; it may mean that asbestos fibers in the lungs trigger something else that causes the cancer. Finally, epidemiologists are hampered by usually small groups, individual resistance to participation and, in a society as mobile as this one, the difficulty of access to good records.

Animal experiments, generally with rodents, have for the most part been very effective in determining whether a given chemical is carcinogenic. But there are many scientists who argue that applying such data to humans, which is called extrapolation, can be misleading. "There is no reason to believe that what causes cancer in

"There is no reason to believe that what causes cancer in animals will do it in human beings and vice versa," says Dr. Irving J. Selikoff.

We depend on scientists and statisticians, such as E. Cuyler Hammond of the American Cancer Society to apply data from experiments and tell us which poisons the human body can tolerate.

animals will do it in human beings, and vice versa," says Dr. Irving J. Selikoff, director of the Environmental Sciences Laboratory in the Mount Sinai School of Medicine in New York. "If you can't extrapolate from a mouse to a rat, how can you do it from a mouse to a man?" he asks. Dr. Selikoff does not question the value of animal experiments, but only maintains that biological systems and responses vary even between different species of rodents, so precise relationships between rodents and humans are just that much more difficult to establish. Furthermore, a typical animal experiment involves 100 or fewer test animals, yet it costs upward of $250,000.

Since the regulatory agencies are not funded to perform anywhere near the number of necessary experiments (there are about 35,000 different pesticide formulations alone), the chemical manufacturers, themselves, incredibly, assume that respon-

sibility. Testing is either done in-house or, more often, is farmed out to "independent" laboratories. In that chemical and drug companies are in business to sell their wares, not to see them squelched in laboratories for being dangerous to human health, it is not too far fetched to suggest that these labs give their clients' concoctions every reasonable doubt.

In mid-1977, the Environmental Protection Agency and the Food and Drug Administration began joint audits of 100 laboratories. Only three, Industrial Bio-Test Laboratories of Northbrook, Ill., and two other labs doing far less testing, have been found to have deficiencies, according to Fred T. Arnold, chief regulatory analyst for pesticides at the E.P.A. And deficences, he adds, are not necessarily deliberately caused. Until the aftermath of that audit, Industrial Bio-Test was one of the biggest of the "independents" and did about 4,300 pesticide tests for such companies as American Cyanamid, Chevron, Ciba-Geigy, Dow, Gulf, 3M, Monsanto, Shell and Velsicol, plus the United States Department of Agriculture and the F.D.A. itself.

According to Mr. Arnold, his agency has no direct authority to regulate laboratories, but it can require that manufacturers undertake "stringent" testing of pesticides by whatever means are necessary. If the tests are judged inadequate or distorted, the E.P.A. can order them repeated, which is what is now happening in the Industrial Bio-Test case. But, Mr. Arnold says, what with redosing and reanalyzing, it will be two years before the various toxicity levels are fully determined. Meanwhile, some of the 123 pesticides in question, among them several of the most heavily used in the world, remain in mass production.

□

Toxic chemicals, including those that can cause cancer, are supposed to be regulated directly by four Federal agencies:

The Environmental Protection Agency, the largest of the regulators, which is responsible for protecting the overall environment.

The Food and Drug Administration, operating under the Department of Health, Education and Welfare, which is responsible for drugs, food, cosmetics, medical devices, television sets and microwave ovens.

The Consumer Product Safety Commission, an independent agency, which monitors such consumer items as children's clothing, patching compound, and aerosol sprays.

The Occupational Safety and Health Administration (OSHA), part of the Department of Labor, which tries to control "hazards in the work-place," from asbestos particles in the air to the number of fire exits.

Although these agencies are required by law to do the actual regulating, they are merely the most visible part of a much larger mechanism. In addition to coming from industry and universities, many of the data used for decision making come from the National Cancer Institute, the National Institute of Environmental Health Science and the National Institute for Occupational Safety and Health. Then there are all of the other agencies, large and small, with primary responsibilities elsewhere which nonetheless become involved in toxic substances research and regulation.

In August 1977, the four main regulatory agencies spawned something called the Interagency Regulatory Liaison Group (I.R.L.G.), which is supposed to facilitate cooperation among them. An I.R.L.G. risk-assessment group announced in February, 1979, that it had come up with methods for deciding whether something is carcinogenic and, if so, on the extent of the risk it poses. The report was then submitted for peer review and public comment.

□

The situation has been so confused that when the E.P.A. recently asked the National Academy of Sciences to evaluate two pesticides—heptachlor and chlordane—one academy committee came up with a risk assessment, while another reported that it hadn't enough information to do so with reliability. The discrepancy came less from disagreement over the validity of extrapolating animal test data than from ambiguities in the law.

What is acceptable and what is not is terribly complicated by the 12 laws, each of which came out of a different Congressional committee, under which the four major agencies function. The Clean Air Act, Water Pollution Control Act, Safe Drinking Water Act and Federal Insecticide, Fungicide and Rodenticide Act, all administered by the E.P.A., permit cost-benefit analysis for setting acceptable levels, but do not require it. The Food, Drug and Cosmetic Act (F.D.C.A.), demands cost-benefit analysis in some instances and prohibits it in others. The F.D.C.A.'s controversial "Delaney clause" forbids carcinogenic additives from being put into food, but since PCB's (as one example) are environmental rather than an additive, they are not covered by the clause.

The elasticity of the Clean Air Act is likely to have prompted Charles L. Schultze, chairman of the Council of Economic Advisers, to warn E.P.A. Administrator Douglas Costle that his agency's broad new proposals to control air pollution would "impose substantially higher costs on business than is necessary." The E.P.A., bending to

A group of asbestos workers demonstrated for a medical clause in their contract to protect them against job related diseases.

such pressure, weakened a proposed smog regulation for the first time in its history. More recently, a memorandum from Mr. Schultze and Alfred E. Kahn, the President's chief adviser on inflation, said that E.P.A.'s pending water-pollution regulations could be "prohibitively expensive" and that the whole program should be re-examined to relate the costs to the benefits. Here, laid bare, is the nub of the antiregulatory argument and the basis for most of the courtroom action: Following all of the rules and regulations is economically crippling and dangerously inflationary.

It is Mr. Schultze's use of the word "necessary" in the air-pollution warning that would send economists and ecologists going for one another's throats. How necessary is clean air and how clean should it be in order to be acceptable? What if requirements for super-duper clean air so constrict business that companies collapse and thousands are thrown out of work? Would those

unemployed workers rather have perfect air, inflation and no jobs, or not-so-perfect air, a healthy economy and work? This is the essence of industry's argument.

Although existing law stipulates that 5 parts per million is the allowable level in fish, as Leon Hetling told Carolyn Cunningham, the F.D.A. in fact proposed in April 1977 to lower that level to 2 parts per million, as Canada had already done. As required by law, the proposal appeared in the Federal Register. Interested parties were given 30 days to file objections and request a hearing.

Robert G. Martin, assistant vice president of the Sport Fishing Institute, an organization representing the multibillion-dollar sport-fishing industry, maintains that he has yet to see studies "implicating PCB's with human health. You really can't be against public health," he says, "but I find it difficult to see where human health has been involved." In checking his files, Mr. Martin adds, "I could find no reference to the institute officially commenting on a proposal to lower allowable limits of PCB's from 5 to 2 parts per million." F.D.A. records, however, tell a different story. They show that Gilbert C. Radonski, executive secretary of the institute, sent a letter to the agency on May 23, 1977, which said that it "has not been demonstrated that an eminent health hazard exists that would justify a reduction in the total of PCB's in fish from 5 parts per million to 2 parts per million." The letter also called for more research on PCB's and for the issuance of "guidelines" on the best ways to prepare and cook fish contaminated by the chemicals. (The institute's motto is: "The Quality of Fishing Reflects the Quality of Living.")

□

Other fishing organizations, such as the Midwest Federated Fisheries Council and the National Fisheries Institute, and even the New York State Legislature, came out for delaying a lowering of the PCB tolerance level in fish "pending further studies on the drastic and possibly tragic economic implications any change would instigate." This position prompted one member of the New York PCB Settlement Advisory Committee to say privately, "What they want, at bottom, is a body count [before they believe that PCB's are carcinogenic to humans]." Meanwhile, the proposal to drop the PCB level, a classic victim of cost-benefit analysis, seems economically unacceptable and is now in limbo.

After years of notable victories by environmentalists, the economics-versus-ecology battle looks as though it is beginning to tilt the other way, in part because of persistent inflation and also because of an equally persistent mistrust of big government and its sluggish and meddling bureaucracy. Some observers even sense a growing backlash among citizens who feel that good old American enterprise is choking to death on hopelessly snarled red tape and ridiculous regulations. They feel, and with some justification, that they are overregulated and underprotected. (Given the President's announcement that OSHA was eliminating 928 "unnecessary" regulations, this sentiment seems at least partly shared by the Administration.)

But because of the long latency period, it is vital to work now to prevent as much cancer as possible in the year 2000, warns Dr. Irving Selikoff of Mount Sinai. "How much risk can you live with? That's the nature of the debate over the word 'acceptable,' and it must be decided by society, not by scientists."

There are about 100,000 known natural chemical entities, 1 to 5 percent of which are thought to be carcinogenic, according to Dr. W. Gary Flamm, a toxicologist in the Food and Drug Administration's Bureau of Foods. So, what do toxicologists eat?

"Everything," he says. "My philosophy is to spread the risk. If I ate one thing [in excess] I'd be in trouble."

Most grains, including corn, and peanuts develop a mold called *Aspergillus flavus* which, in turn, produces aflatoxin B, a known carcinogen. Peanut butter contains an almost infinitesimal level of aflatoxin B, says Dr. Flamm. But he adds, "This doesn't mean that you shouldn't eat peanut butter. All it means is that you'd be smart not to eat peanut butter, or anything else, exclusively."

New Battle Over Endangered Species

by Philip Shabecoff

The Furbish lousewort, despite its unprepossessing name, is not without a certain modest charm when it first emerges in the springtime along the steeply sloping banks of the St. John River. Its leaves are delicate and fernlike, its stem is slender and, in due season, it diffidently puts forth two or three pale yellow blossoms. By the time the boreal forest of northern Maine has donned its brilliant autumn foliage, however, the lousewort, a distant relative of the snapdragon, has turned into a brown, blowzy, utterly inelegant growth. Its leaves are curled and brittle, its stalk is withered and its seed pod, gaping and empty, droops like a tired slattern. Standing among clumps of vivid red fireweed, the aging lousewort is all but invisible.

Yet this little weed created a furor when it was reported that the plant might block construction of two massive dams planned for a $700 million hydroelectric project bigger than the Aswan Dam in Egypt. Widely believed to be extinct, the little-known lousewort was spotted growing in the path of the proposed Dickey-Lincoln power project in the summer of 1976. Because it is in danger, the lousewort has been placed on the Endangered Species List maintained by the Fish and Wildlife Service of the Department of the Interior. Long before the discovery, however, construction of the dams had been under attack by environmentalists and sportsmen for other reasons.

Similarly, after local conservationists in February 1976 brought a lawsuit to halt further work on the Tellico Dam, which is part of a $119-million regional development project of the Tennessee Valley Authority, the Little Tennessee River was found to be the "critical habitat" of the snail darter. Conservationists found that the three-inch fish, an endangered species, lived in the last 38 miles of clear, free-flowing water on the Little Tennessee River; it cannot live in the still water that would be created by the dam impoundment.

Because of the growing concern that some obscure species of life may block multimillion-dollar projects, Congress has before it several amendments to the Endangered Species Act of 1973 that sponsors say would make the act "more flexible." The act, which restricts only projects involving the Federal Government, was set up to prevent the elimination of species, subspecies and population segments of plants and animals; the amendments would enable some projects to proceed even if certain species are made extinct in the process.

NEW BATTLES OVER ENDANGERED SPECIES

Environmentalists have regarded the 1973 act as a first line of defense for the growing number of animals and plants facing extinction. In the course of safeguarding the life of these species, however, conservationists have come up against unexpected dilemmas. The Furbish lousewort is a case in point. It was seen for the first time in 30 years in 1976, when Dr. Charles D. Richards, a botanist from the University of Maine who was working on an environmental-impact statement for the Corps of Engineers, came across a clump of the weed along a stretch of the river bank. The Corps of Engineers has long had plans to construct a pair of dams on the St. John. The Dickey-Lincoln hydroelectric project would flood much of the river valley and, in so doing, destroy half of the 800 louseworts on the river banks.

To those fighting the Dickey-Lincoln project, the plight of the Furbish lousewort poses a particularly sticky problem. While they are against any deliberate attempt to extinguish a species, they recognize that the weed is being turned into something of a political scapegoat by proponents of the dam. "The lousewort is actually damaging to our efforts to save the river because people think it is ridiculous that a weed should hold up a project costing millions of dollars," said Wayne Cobb, former assistant director of the nonprofit Natural Resources Council of Maine, which is leading the battle to save the St. John from drowning.

"There are a lot of good reasons for not building those dams," said Mr. Cobb. He explained that the dams would destroy one of the last wild rivers in the East and eliminate many miles of some of the best white-water canoeing anywhere. It would submerge 88,000 acres of valuable timberland while supplying virtually no power to impoverished northern Maine because most of the electricity that would be generated is destined for the Boston area.

"If you look at the energy crisis and the economics of the dam," says E. Lee Rogers, attorney for the council, "you realize Dickey-Lincoln is not the right answer — there really are better alternatives. For instance, estimates are that the dam would eventually cost $1 billion. A study by the Audubon Society shows that if you were to take the $1 billion and insulate the buildings in New England, you'd have a much greater energy savings."

It appears now that if the dam is blocked it will not be by the Furbish lousewort. Keith M. Schreiner, manager of the endangered species program for the Fish and Wildlife Service, says that official consultations with the Army Corps of Engineers have yet to begin. But he feels that there are a number of ways the issue might be resolved: "It's possible you could establish a population upstream from the impounded area. Or you could arrange for the full protection of the Furbish lousewort populations that already exist below the dam as well as in Canada."

It is the case of the Tellico Dam and the snail darter that has officials stymied. Attempts at

The Little Tennessee River was found to be the critical habitat of the snail darter and a lawsuit brought on behalf of the 3-inch fish halted further work on the Tellico Dam.

A plan to channelize the Little Auglaize River in Ohio was altered because it would have destroyed the tree canopy on either side of the river that is used by the Indiana bat for food and living quarters.

transplanting several hundred fish into another river have been made and, according to the T.V.A., have thus far been successful. But transplanting an animal population is simply not as easy as transplanting a plant population. "One of the hardest things in wildlife management is to establish a viable breeding population of animals in a new habitat," says Mr. Schreiner. "For every 100 attempts, there are 95 failures." He adds that it takes years before a population can be considered successfully established.

Without a compromise, matters quickly came to a head. The howl raised by Congressmen from Tennessee and other supporters of the dam led to Senate subcommittee hearings on the Endangered Species Act amendments. (The case of the Tellico Dam vs. the snail darter is also before the Supreme Court.)

Partisans of the dam argue that it would provide flood control, recreational and industrial development and more power. They also assert that

it would be ridiculous not to use a dam that is already almost entirely completed at a cost of $117 million. Defenders of the snail darter, on the other hand, are making a strong case that the dam is a pork-barrel scheme that would benefit relatively few people. They also argue that the Little Tennessee should be saved as the last free-flowing portion of the already abundantly dammed Tennessee River system.

☐

At the Senate subcommittee hearing, witnesses for the Carter Administration, which backs a strong endangered-species law, pointed out that hundreds of conflicts between construction projects and endangered species have been resolved through compromise solutions and only three cases have gone to court. One concerned the Meramec Park Dam in Missouri and the Indiana bat. The Fish and Wildlife Service indicated that

The expansion of the Los Angeles airport is being blocked because the El Segundo blue butterfly lives in a few acres which happen to be at the end of the airport's runway.

the dam would jeopardize the bat's existence, but a District Court ruled in favor of the completion of the dam, claiming that the entire species was not jeopardized.

Nevertheless, many Congressmen continue to feel the act is in need of revision, and several amendments have been proposed. The one considered to have the best chance of enactment is sponsored by Senator John Culver, Democrat of Iowa and chairman of the subcommittee that held

the hearings, and Senator Howard H. Baker of Tennessee, the Republican Senate leader. The amendment would create an interagency review board of high-level-Government officials who could permit a Federal project to destroy a species of life if the project benefits "clearly outweigh" the value of the species. Senator Culver explains that this action will stave off other, more drastic amendments that would seriously cripple the Endangered Species Act. Not everyone agrees.

144

"It is just not necessary to amend the act," says Mr. Schreiner of the Fish and Wildlife Service. "Tellico Dam has been *the* only absolute impasse. In other cases, some compromise has been worked out." He and others who testified before the subcommittee worry that, if the amendment is passed, the Federal Government will be placed in the position of having to consciously extinguish a harmless species. Michael Bean of the Environmental Defense Fund, who called the 1973 act "a determined attempt to keep the concept of the biblical ark afloat," commented: "Along the way, it is true that a lot of species have fallen off the ark. Some have even been unwittingly crowded off by man himself, but never before has any species been intentionally thrown overboard."

In 1973 there were 418 species listed as endangered. Today, there are 662 life forms being protected and another 2,000 being reviewed for possible inclusion on the list. As each species is added to the endangered list, the prospects increase for conflict between the right of plant or animal to exist and the demands made by humans on the land and its resources.

A Federal dredging operation was halted to protect the critical habitat of the Higgin's eye pearly mussel.

The El Segundo blue butterfly, which lives in only a few square acres which happen to be at the end of the runway of the Los Angeles airport, is blocking expansion of the airport.

A biological opinion given indicated that the Columbia Dam in Tennessee would jeopardize the existence of various endangered mussels and snails. Local representatives and senators have introduced legislation to exempt the dam from the Endangered Species Act.

Federal officials denied a permit to the Florida Power and Light Company to build a nuclear power plant in an undeveloped 10,000-acre coastal site in southern Dade County, near Miami. The power plant was to occupy a few hundred acres, with the rest of the land kept wild, but that tract was in the critical habitat of the manatee, an endangered species of aquatic mammal. The irony in this case is that the power company would probably have to sell the land, and the most likely buyer would be a real-estate developer who, unaffected

The manatee, an endangered species of aquatic mammal, living along the coast of Florida near Miami.

by the endangered-species law, could drain the area for private housing. In the view of many conservationists, the Endangered Species Act is not strong enough in that it restricts only projects in which the Federal Government is involved.

Extinction has been a common fate for almost all species that have appeared on earth. Sooner or later, even the most powerful, dominant forms, such as dinosaurs, pass from the scene. But in modern times, human activity has speeded up the pace at which species vanish at an alarming rate. In prehistoric times, it is believed, one species became extinct every 10,000 years. Around the year 1600, the rate was about one every thousand years. Now, estimates of the extinction rate range from one a year to 20 a year, although the latter figure is regarded as high. One environmentalist figures there are now one million species, subspecies and general populations in trouble. "A lot of them will be extinct before we even know they exist," he says.

The culprit in this acceleration of "unnatural extinction" is man. Hunting, particularly commercial hunting, is one way in which humans have wiped out species. Far more frightening is the rapidly increasing toxification of the earth by industrial society. The Eastern peregrine falcon was wiped out by the pesticide DDT, which made the bird of prey's eggs too fragile to hatch. As more and more poison seeps into the air, the land, the lakes, streams, rivers and oceans, the number of endangered species will swell rapidly.

But by far the most devastating impact on wild animals and plants comes from the destruction of their natural habitats by human activity. As forests are leveled, wetlands drained, water tables lowered and natural ecosystems paved over for roads, dams, shopping centers and suburbs, the elbow room and nutrients required by a broad range of species are rapidly diminishing.

"The primary cause of extinction is not bad hunters shooting defenseless animals," says Fish and Wildlife's Keith Schreiner. "It is the change and destruction of habitat. Man's ability to change the earth has increased 20-fold over the past 50 years with the use of bulldozers and cranes and other earth-moving equipment. If enough people care, we can slow the process. But common sense tells me the future is bleak."

So bleak is the picture, in fact, that the bulldozer and not the atomic bomb may turn out to be the most destructive invention of the 20th century. The protection of critical habitats is the most pub-

The eastern peregrine falcon was destroyed by the pesticide DDT which made its eggs too fragile to hatch.

licized aspect of the work of the Fish and Wildlife Service, in large part because of the brawl between the snail darter and the Tellico Dam. Sometimes the Government acquires land to give a species room to exist. Several years ago, for example, it bought land in the Florida keys to provide *Lebensraum* for the vanishing key deer. More recently, the Federal Highway Administration agreed to buy 2,000 acres around the site of a proposed highway interchange, and turn them over to the Fish and Wildlife Service, thus preserving the critical habitat of the Mississippi sandhill crane.

Perhaps the most famous effort to save an endangered species involves the whooping crane, a stately wading bird with a haunting cry. Hunted for its beautiful feathers, the bird had all but vanished by the early 1940's, with only about 15 left in the wild. Careful nurturing by the wildlife's research center on the Patuxent River in Maryland, and in other such centers, kept the species going. In 1978 there were about 100 whoopers alive and the species has moved a few steps back from the edge of extinction. Nine areas in seven states have been set aside as critical habitats. And 12 whooper eggs placed in sandhill-crane nests at Grays Lake National Wildlife Refuge in Idaho hatched.

The Federal Highway Administration agreed to buy 2,000 acres around the site of a proposed highway interchange, and turn them over to the Fish and Wildlife Service, thus preserving the critical habitat of the Mississippi sandhill crane.

Arguments for the effort and cost of protecting species are often couched in esthetic and ethical terms such as the beauty of diversity or the arrogance of man in assuming that the earth was created solely for his benefit. "There is probably nothing more criminal than destroying a life form," says Mr. Schreiner. "You are not talking about a single life but an entire species. Once a life form is gone you can never bring it back. What a tragedy!"

One problem that is viewed with growing seri-

ousness is the reduction of the earth's aggregate gene pool with the passing of each species. By allowing plants and animals to pass out of existence, it is argued, we may be losing things of incalculable and irreplaceable value to man, if we only knew it. Who knew the value of bread mold 50 years ago? But the discovery of penicillin has saved millions of lives.

Other discoveries — such as the use of the blood of horseshoe crabs as a detector of toxins in intravenous fluids, or the chemical in a plant that was used to develop birth-control pills — suggest that "worthless" creatures and weeds may be very valuable indeed. "We don't actually know that the Furbish lousewort is not useful," says botanist Charles Richards when asked why the little weed should be saved. "Perhaps it could be used as a part of a gene pool for breeding. Perhaps it might have some medicinal value we haven't discovered yet."

Increasingly, scientists and environmentalists are coming to the conclusion that the destruction of species is directly related to the viability of the human race itself. The disappearing life forms are likened to the canaries that coal miners used to carry down into the shafts before the advent of the safety lamp. If the canary died, it was an indication that poison coal gas was in the vicinity and that the miners themselves were in grave danger.

Today, the endangered species are an early-warning system indicating that the human habitat is becoming dangerous. Dr. George Woodwell, director of the Ecosystems Center at the Marine Biology Laboratory in Woods Hole, Mass., finds the loss of species "very frightening" and says that "it has a direct bearing on and cannot be separated from a larger problem — the biotic impoverishment of the earth, which is reducing the capacity of the environment to produce services. It is one of the great issues of our time, right up there with nuclear proliferation, the stability of government and health care. The ultimate resource is the biota — there is no other. And we are destroying it."

Environmentalists say they are accused of being alarmists. "But when you think of the tropical forests being destroyed," said Mr. Schreiner, "when you see the rampant stream pollution, the soils full of pesticides, animals poisoned to the point where they can't reproduce, well, you've got to believe we are putting ourselves out of business. And that may not be far away."

So ask not for whom the Furbish lousewort lives. It lives for you.

BIRDS AND FISH VS. HIGHWAYS AND DAMS

The primary cause of extinction of plants and animals is the alteration or destruction of their habitats. Protection of the endangered species shown here has led or may lead to blockage of major Federal projects.

HIGGIN'S EYE PEARLY MUSSEL

PEREGRINE FALCON

BROWN PELICAN

GRAY BAT

EL SEGUNDO BLUE BUTTERFLY

RED-COCKADED WOODPECKER

| Dam | Space Shuttle | Highway | River Channeli-zation | Bomb Testing | Airport | Power Plant | Timber Manage-ment |

148

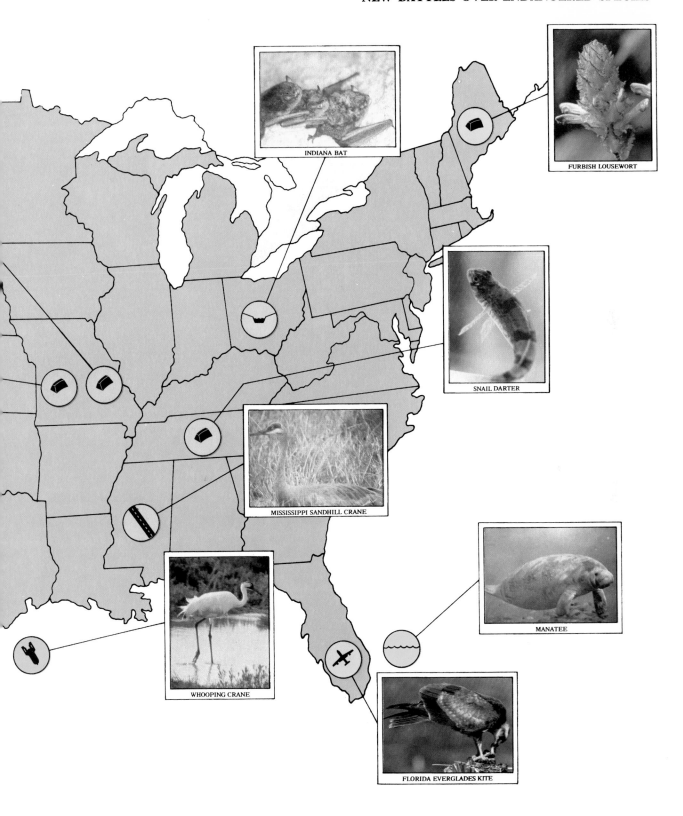

INDIANA BAT

FURBISH LOUSEWORT

SNAIL DARTER

MISSISSIPPI SANDHILL CRANE

MANATEE

WHOOPING CRANE

FLORIDA EVERGLADES KITE

Perhaps the most famous effort to save an endangered species involves the whooping crane which was hunted for its beautiful feathers and was nearly extinct by the early 1940's.

BIRDS AND FISH VS. HIGHWAYS AND DAMS

In addition to the species cited in the article above, other flora and fauna have blocked major projects. They include:

Brown pelican and peregrine falcon:

The landing route planned for the Space Shuttle north of Los Angeles in the Mojave Desert is of concern to the Fish and Wildlife Service. In the flight path of the shuttle are the Anacapa Islands — part of the Channel Islands — on which live one of the largest colonies of brown pelicans on the West Coast; the Channel Islands had been the nesting sites of peregrine falcons. Sonic booms caused by the Space Shuttle could disrupt the brown pelicans during the nesting season.

Red-cockaded woodpecker: These birds make old trees their habitat. In its timber-management plan for the national forests in Texas, the Forest Service would have removed the trees considered fire hazards — i.e., the very trees that woodpeckers live in. The Forest Service has accepted recommendations to modify that plan.

Whooping crane: The U.S. Air Force has agreed to restrict flights over Matagorda Island, Tex., and to eliminate bomb tests during the cranes' winter residence on the island.

Gray bat: Construction of the Harry S. Truman Dam, Mo., would have destroyed the caves in which the bats live. The Army Corps of Engineers is now modifying its plans so as to include the fencing of cave entrances.

Indiana bat: A plan to channelize the Little Auglaize River, Ohio, would destroy the tree canopy on either side of the river that is used by the bats for food and living quarters. The Soil Conservation Service is altering its plan.

Florida Everglades kite: Forty birds, or 25 percent of the world's Florida Everglades kite population, nest in a willow island in the Florida Everglades. The island would be directly in the path of aircraft to and from the proposed Miami Replacement Jetport, Fla. Such flights would prevent the birds from nesting. The Federal Aviation Administration is restudying the location of the airport.

The Promise and Perils of Petrochemicals

by Barry Commoner

The ancient goal of alchemy, turning baser metals into gold, can now be achieved by modern atom-smashing techniques, but the expense is prohibitive. However, where classical alchemy has failed to meet the test of the marketplace, that elegant new alchemy of our times—the petrochemical industry—has succeeded.

The petrochemical industry's products, made chiefly out of crude oil and natural gas, make up a marvelous catalogue of useful materials: cloth with the sheen of silk or the fuzziness of wool; cables stronger than steel; synthetics with the elasticity of rubber, the flexibility of leather, the lightness of paper, or the workableness of wood; detergents that wash as well as soap without curdling in hard water; chemicals that can kill dandelions, but not grass; repel mosquitoes, but not people; diminish sniffles, reduce blood pressure, or cure tuberculosis.

But something has gone wrong. Increasingly, the chemist succeeds, brilliantly, in synthesizing a new, useful, highly competitive substance, only to have it cast aside because of its biological hazards: Food dyes and fire-retardants for children's sleepwear are banned because they may cause cancer; a new industry to produce plastic soda bottles, developed at a cost of $50 million, comes to an abrupt halt as the Food and Drug Administration discovers that a chemical which may leach out of the bottles causes tumors in mice; pesticides are taken off the market because they kill fish and wildlife; firemen would like to ban plastic building materials because they produce toxic fumes when they burn.

The petrochemical industry, by any ordinary criteria—its rate of growth, its profitability, the apparent eagerness of consumers to purchase huge quantities of its products—is the most successful in postwar America. Yet it is now caught by forces that threaten its viability. The industry's raw materials, petroleum and natural gas—once cheap and plentiful—are now becoming astronomically expensive as the supplies shrink and become increasingly uncertain. The industry has been beset by Federal regulations and by pressure from the public, often based on the widespread fear of cancer, to control not only the dissemination of its toxic wastes, but also of some of the industry's most salable products, such as saccharin. These pressures on a major American industry have important implications for the economy, for the availability of jobs, for the future of industrial technology, and will affect the availability of products that most of us have taken for granted, and may soon have to do without.

THE PARADOX OF MODERN TECHNOLOGY

The petrochemical industry is a huge success in the marketplace. Since World War II, the

"Kapton" polymide film made by Du Pont is shown here being applied to a conductor. This film, derived from petrochemicals, is extremely strong and is used for insulation when resistance to high heat and extreme cold are necessary.

152

industry's detergents have driven soap out of a market that it monopolized for perhaps a thousand years; in textiles, synthetic fibers have massively displaced cotton and wool; plastics have replaced long-established uses of metals, wood and glass; food production has become heavily dependent on fertilizers, pesticides and other agricultural chemicals; synthetic drugs and toiletries have become a major enterprise in an area of commerce once represented by concoctions of herbs.

In most industrial nations the petrochemical industry has become the fastest-growing sector of manufacturing. In the United States, in the last 30 years, it has grown at an annual rate of about 8 percent, twice the rate of growth of manufacturing industries as a whole. The industry's economic rewards have been correspondingly large: In that period its rate of profit has generally led those of all major industries; and the pharmaceuticals have led all the rest, recording profits ranging about 20 percent, as compared with the 11 percent average for manufacturing as a whole.

Yet the petrochemical industry dramatizes the paradox of modern technology: Its blessings are mixed with plagues. Here are some of the more recent troubles:

● In Michigan, because a chemical plant inadvertently combined a fire-retardant with animal feed, thousands of cattle and millions of chickens were lost, and hundreds of people made ill.

● In Virginia, because of the careless operation of a small insecticide plant, the entire Chesapeake Bay fishing industry has been threatened, and dozens of chemical workers have had their health ruined.

● In Seveso, Italy—a small town near Milan—the accidental release of a few pounds of a highly toxic chemical from a petrochemical plant has caused serious illness among hundreds of children and forced the evacuation, now in its second year, of 700 people from their homes.

The most chilling prospect is that much of the cancer problem in the United States may eventually be laid at the door of the petrochemical industry. The most drastic actions taken against the petrochemical industry thus far— the banning of DDT and other chlorinated hydrocarbon insecticides, and most recently PCB—have been largely based on their hazards

as possible carcinogens. It has been estimated that perhaps three-fourths of the incidence of cancer in the United States is due to environmental agents, and a recent county-by-county survey of cancer incidence shows a significant correlation with the local concentrations of petrochemical operations. The highest incidence of bladder cancer in the United States is found in Salem County, N.J., an area dense with refineries and petrochemical plants. A particularly alarming fact is that since the 1960's the downward trend in the death rate among United States males has stopped and reversed itself, and according to a recent U.S. Public Health Service survey, cancers (other than those attributed to smoking) are prominent among the causes of the rising death rate. Since many cancers may develop only 10 years or more after exposure to the instigating substance, and since widespread exposure to petrochemical products has occurred only in the last 20 to 25 years, there is reason to fear that these changes may foreshadow an upsurge in the incidence of cancer. If so, we may be facing an epidemic of environmental cancer, induced by past, irremediable exposures to petrochemicals.

The quandary of the petrochemical industry is that the unique reasons for its success and growth—that it can produce a growing variety of new man-made substances, and can sell them cheaply, but only in very large amounts—are

When a chemical plant combined a fire-retardant with animal feed, thousands of cattle were lost and hundreds of people became ill.

This bottle is coated with "Surlyn," an ionomer-coating made through petrochemical research. Dropped from four feet, it retains 95 percent of its glass fragments resulting in a safer bottle for consumers.

themselves the sources of its growing threat to society. Pressed by its economic structure to create ever more chemically complex man-made products on huge scales, the industry now confronts a hard fact of nature—that the more complex these products, the more likely they are to harm living things, including people, and the more widespread they are, the greater their toxic impact.

HUNDREDS OF SUBSTANCES

The petrochemical industry is fundamentally different from all other production industries. To begin with, it is a *process* industry. This sets it off from ordinary manufacturing such as the production of automobiles, dresses or frying pans. In these enterprises, the *form* of the final product is different from that of the starting materials, but the substances of which it is made—steel, glass and rubber; cotton, wool or nylon; iron or aluminum—remain unchanged. In contrast, in a process industry a new *substance* is produced: Iron ore, coke and oxygen, properly combined, become steel; sand and lime, heated together, become glass; two petrochemicals, butadiene and styrene, appropriately reacted, become synthetic rubber.

However, where the other process industries produce only a few different types of steel or glass, the petrochemical industry converts oil and natural gas into many hundreds of chemically different substances. The industry is based on the chemistry of the carbon atom, which can combine in enormously variable ways with the other atoms that most frequently occur in carbon-containing compounds, hydrogen, oxygen and nitrogen. The industry's processes branch like some great genealogical tree: From crude oil, which contains 10 major primary constituents, the industry produces about 75 chief intermediate chemicals, which are converted into about 100 large-scale end products, each manufactured in amounts ranging from one million to three billion pounds per year. At each successive stage, the industry produces more numerous, more varied, and, as we shall see, more dangerous substances.

Materials flow constantly through the industry's gargantuan network of multiply interconnected pipelines, separators and reactors. Since storage is very expensive and therefore limited

The conductors in this communications cable are surrounded by a core wrap of "Mylar" polyester film. Mylar is used for magnetic tapes, microfilm and packaging as well.

"Vexar" sleeving stretches to fit and taper over all kinds of shapes. This is one more variety of petrochemical product that has been developed in recent years.

in capacity, everything that enters the system has to go somewhere: It is either carted off as a final product, burned by the industry itself as a fuel (or wastefully as a "flare") or—too often—expelled into the environment as waste.

The petrochemical industry has a strong, built-in tendency to proliferate, to elaborate the number and variety of its products. Consider what happened when the industry began to market one of its most successful products, polyethylene film—that ubiquitous wrapping around everything from supermarket beefsteaks to kitchen leftovers. Through the strange, Rube

Goldberg economics of the petrochemical industry, acrylic fiber captured the rug market largely because supermarkets began to use huge amounts of polyethylene film to wrap vegetables and cuts of meat. It happened this way: The film is manufactured from ethylene. Ethylene production yields, as a major byproduct, propylene. At first, this was burned by the industry as a fuel. But then it was discovered that propylene could be converted to acrylonitrile, which could then be used to manufacture acrylic fibers for rugs. When propylene was sold for that purpose, it brought a price twice its value as fuel.

This saving significantly reduced the cost of producing ethylene, ultimately cutting the price of polyethylene film as well, and expanding its sales. Therefore, at the same time, acrylic fiber was marketable at an attractive price because it could be produced cheaply and in large amounts.

Thus, new products are often created not so much to meet the consumer's needs as the industry's. This approach is clearly reflected in a typical petrochemical company's (the Hooker Chemical Company) research policy: "Rather than manufacturing known products by a known method for a known market . . . the research department is now free to develop any product that looks promising. If there is not a market for it, the sales development group seeks to create one."

This helps to explain why petrochemical products characteristically *invade* the rest of the production system. Since synthetic detergents appeared in the 1940's, they have captured about 85 percent of the market once held by soap. Since 1950, synthetic fibers have taken over about 70 percent of the United States textile market, once largely held by cotton and wool. The production of plastics has grown at the rate of 16 percent per year, while the production of competitive materials has been much slower: Leather has increased at an annual rate of only 1.2 percent, paper at 4.8 percent, and lumber production has decreased at a rate of 0.5 percent per year.

ECONOMIC IMPERIALISM

The petrochemical industry seems to have developed a kind of economic imperialism, forcing consumers to give up old products, most of them natural, for synthetic replacements. If this judgment seems harsh, it is nevertheless shared by one of the leaders of the British petrochemical industry. Lord Beeching: "Instead of producing known products to satisfy existing industrial needs, it [the petrochemical industry] is, increasingly, producing new forms of matter which not only replace the materials used by existing industries, but which cause extension and modification of those industries. . . . To an increasing degree it forces existing industries to adapt themselves to use its products." The truth of Lord Beeching's gener-

alization is evident to anyone who has recently tried to find a pure cotton shirt, a laundry cleaner free of synthetics, or a wooden clothespin.

The rapidly changing energy situation is likely to intensify these problems. Fearing a shortage of natural gas, a large sector of the petrochemical industry is planning to substitute crude oil as a source of ethylene. This switch will demand new capital investments that are three to four times greater than current investments in natural-gas-based facilities, no small feat in a period of capital shortage. Using crude oil for ethylene production will mean that many more leftover chemicals will be produced, and, according to a recent analysis of the situation in Business Week, "a much wider variety of products." Not surprisingly, Edward G. Jefferson, senior vice president of DuPont, sees this new situation as "one in which the petrochemical industry becomes an even more important mainstay of the United States economy." Once more, the peculiar technological and economic design of the petrochemical industry forces it to proliferate new products and to penetrate more deeply into the national economy.

The petrochemical invasion has been particularly targeted against the largest, most long-established markets—for clothing, building materials, furniture, appliances, cleansers and other necessities. These invasions succeed because synthetic petrochemical products can be manufactured in large volumes at low prices. This capability is, in turn, the outcome of the technological design of the petrochemical industry. The machinery used in the flow processes that characterize petrochemical production is very complex, replete with miles of piping, specially built reactor vessels, numerous valves, switches, recording devices and elaborate controls. The machinery is very expensive; the petrochemical industry is the most capital intensive of all major manufacturing industries, producing only about 80 cents of value added per dollar of capital invested, as compared with about $3.64, in the case of a typical natural competitor, the leather industry. Since the size and the output of a petrochemical plant can be increased a great deal without a proportional increase in the numbers of expensive valves and controls—or in the amount of labor needed to operate them—there is a considerable economy of scale. A large investment in machinery is more profitable than a small one. As a result, large petrochemical plants are much more economically efficient than small ones. If the capacity of a typical petrochemical plant is increased from one million to 10 million pounds per year, the cost of manufacturing a pound of product is reduced by two-thirds. Thus, petrochemical products can be sold at a low price, but only in very large amounts.

This gives the petrochemical industry a powerful economic advantage in invading large, well-developed markets such as the clothing field. These markets usually exhibit what economists call a "high elasticity of demand": Demand for a particular product rises disproportionately with a fall in price, and vice versa. Synthetic fibers and plastics can readily invade such a market by selling at the low prices made possible by the economies of scale and the byproduct development typical of the petrochemical industry. Thus, between 1960 and 1970, as the price of polyvinyl chloride fell by 75 percent, consumption increased by 500 percent, enabling the plastic to replace leather in clothes, handbags, shoes and gloves.

SELF-FULFILLING PROPHECY

For all these reasons, the economic success of the petrochemical industry has been a self-fulfilling prophecy. Once the industry became established, it was bound to grow, and, by further reducing prices, growth carried the industry into new markets, further accelerating its expansion. The industry would appear to be an invincible competitor in the marketplace of a free economy.

Yet the industry has been increasingly vulnerable to the dangers inherent in its own production processes. That the industry itself has thus far been unable to reduce its hazards to a level acceptable to the public is evident from the intensely critical response to recent environmental catastrophes. The Kepone accident, for example, resulted in an unprecedented fine—$13 million—levied against the Allied Chemical Corporation for its responsibility in the disaster; civil suits involved more than $100 million in damage claims.

Additional evidence that the industry itself can no longer cope with the growing risks inherent in its operations is found in its

These injection molded ski boots derive their toughness from an ionomer resin which is a petrochemical product.

insurance record. Ordinarily, commercial enterprises protect themselves against such risks by obtaining insurance to cover the expected cost of possible damages. When the risk and the possible damage are high, the premium is large; when the possible damage is very great, and difficult to compute accurately, insurance is usually unobtainable. In recent years, the petrochemical industry has suffered from both of these difficulties. A recent survey of the problem in Chemical & Engineering News complains of "skyrocketing insurance premiums, with renewals sometimes 30 to 50 times higher than the old rate. Policies canceled, often without notice or reason. Throughout the industry, companies are finding it increasingly difficult to obtain product liability insurance."

Two of the petrochemical industry's trade associations have considered establishing their own insurance companies, somewhere in the Bahamas. Another proposed solution is no-fault insurance, possibly based on a Federal insurance company. Like the nuclear-power industry before it (which, by virtue of the Price-Anderson Act, is now provided with Federal insurance coverage against the huge damage that might result from an accident), the petrochemical industry may find it necessary to shift the burden of insurance risks from its own, private responsibility, to society's.

Meanwhile, some of the industry's earlier economic advantages have backfired. Originally, the industry's raw materials, petroleum and natural gas, were plentiful and available at a constant price. Now the situation is dramatically different. With the energy crisis, the industry's dependence on these fuels has become a serious economic liability, as supplies become shaky and their costs rise at an astronomical rate. At the same time, the petrochemical industry may be particularly vulnerable to the growing shortage of investment capital. The industry's vaunted high labor productivity, which is nearly three times the average for that of all manufacturing industries, is achieved at the expense of very intense investments of capital.

THE DANGERS OF SYNTHETICS

The clash between the economic success of synthetic petrochemicals and their increasing vulnerability to biological complaints is the inevitable result of the very fact that the petrochemicals *are* synthetic—made by man, not nature. In every living cell there is a tightly integrated network of chemical processes which has evolved over three billion years of trial and error. In all of the countless organisms that have lived over this time, and in all of their even more numerous separate cells, there has been a huge number of opportunities for chemical errors—the production of substances that could disrupt the delicately balanced chemistry of the living cell. Like other evolutionary misfits, any organism that made these chemical mistakes perished, so that the genetic tendency to produce the offending substance was eliminated from the line of evolutionary descent. One can imagine that at some point in the course of evolution some unfortunate cell managed to synthesize, let us say, DDT—and became a casualty in the evolutionary struggle to survive.

Another requirement for evolutionary survival is that every substance synthesized by living things must be broken down by them as well—be biodegradable. It is this biological rule which establishes the distinctive closed cycles of ecology. When petrochemical technology synthesizes a new complex substance that is alien to living things, they are likely to lack the enzymes needed to degrade the substance—which then accumulates as waste. This explains why our beaches have become blanketed in debris, since nondegradable synthetics have replaced hemp cordage, wooden spoons and paper cups, which, because they were made of natural cellulose, soon decayed.

The likelihood that a synthetic organic chemical will be biologically hazardous increases with its complexity; the more elaborate its structure, the more likely that some part of it will be incompatible with the normal chemistry of life. While only 10 percent of the relatively simple primary constituents of petroleum and natural gas are classified as "highly toxic," by the time the petrochemical tree has branched several times to yield much more complex products, the proportion of highly toxic substances has increased to about 50 percent. Thus, as the petrochemical industry is forced by the logic of its economic design to produce increasing numbers of ever more complex synthetic substances, it also increases the risk that some of these products will be dangerous to living things.

As the uses of a product expand, it comes in contact with people in rapidly increasing numbers and in new ways. Then, the built-in propensity of man-made organic chemicals to cause biological troubles has many new opportunities to express itself. Unfortunately, the industry does not readily respond to this change. According to a recent Environmental Protection Agency survey, as production expands ". . . there is essentially no mechanism that triggers expanded toxicological and environmental testing. Such expanded testing, when it occurs, is nearly always in response to a reaction to some adverse finding outside the company." The survey points out, for example, that although United States production of vinyl chloride—a substance recently shown to cause cancer—increased from 321 million pounds in 1952 to 5,300 million pounds in 1973, the industry's assessment of the health hazard failed to keep up with the sharply increased human exposure "until after some adverse publications in the scientific literature," as the E.P.A. report says.

Nevertheless, whether heeded or not, the

earliest warning that a petrochemical has a potential for large-scale disaster usually comes from *within* the industry: Chemical workers, who inevitably are most exposed to the substance, become ill. For example, production of PCB expanded from 20,000 tons in 1960 to 43,000 tons in 1970, although as early as 1941 the severe effects of PCB on workers' health had caused public-health officials to publish warnings that "human contact with PCB should be avoided." Had this admonition been heeded, the expanded use of PCB, especially in products accessible to human contact, such as plastics, paint, ink and paper, and its spread into the environment—not to speak of repeated industrial exposures—could have been avoided.

Such failures to appreciate in good time the biological hazards of petrochemicals are sometimes not so much a matter of neglect as of impotence. In some cases, analytical methods are unable to detect the very low levels at which many substances are toxic. For example, PCB was detectable in the environment only after chemical analysts learned how to separate it from its close relative, DDT. In other cases, as synthetic substances penetrate into the intricate network of the environment, they begin to interact, not only with natural substances but with each other, creating a statistical nightmare. Thus, many of the several hundred synthetic organic chemicals that now find their way into water supplies readily react with the chlorine used in water purification and form new substances, which may in turn react with each other. A common product of these reactions is chloroform, which has now become the most ubiquitous synthetic organic pollutant of United States water supplies; it is a carcinogen. In the area of Charleston, W. Va., where seven large petrochemical plants and several smaller ones annually emit tons of waste substances into the air, 128 different synthetic organic compounds have been identified among them. Several of the waste substances can cause cancer and birth defects. No one knows how many other hazardous substances are produced by the further reaction that may occur among the original pollutants, once they mix in the air.

Enclosures like these "Lucite" sheets provide security and an open appearance in this bank. But such sheets are not bio-degradable.

THREAT TO THE INDUSTRY

The most serious threat to the petrochemical industry is that society is becoming less willing to bear the "externalities"—the risks that the industry has imposed on society as a result of its private success. The strongest evidence is the recent passage of the Toxic Substances Control Act late in 1976. The act empowers the administrator of the Environmental Protection Agency to regulate the manufacture and distribution of chemical substances judged to "present an unreasonable risk of injury to health or the environment." In reaching this judgment, the administrator must conduct open hearings in which all interested parties, including public-interest and citizens' groups, can participate with Federal financial assistance if needed. In establishing a rule to govern a particular substance, the administrator must consider not only its hazard to health and the environment, but also its benefits, the availability of substitutes which are less hazardous, the "reasonably ascertainable economic conse-

The use of plastics in automobile parts eliminates steps in manufacturing. For instance, this bumper strip replaces the old nuts and bolts with a built-in tab insert.

quences of the rule . . ." and its "social impact."

Given this broad mandate for an open discussion of a substance's biological hazards and its economic and social virtues and faults as compared with those of alternatives, this new legislation may result in two very different outcomes for the petrochemical industry and for society.

One possible outcome is that enforcement of the act will become bogged down in niggling debates over what constitutes "an unreasonable risk of injury" and over the comparable benefits of each particular substance. Predictably, industry representatives will argue that the risk of injury is insignificant, while public-interest groups will argue the opposite. The results will be debated; analytical methods questioned; statistics compared. In view of the enormous number of substances to be considered, running into the thousands, it is easy to envisage the end result of this approach. Substance by substance, disputed judgments of their hazards will slowly emerge from a vast bureaucratic jungle. It is to be hoped that this will gradually reduce the number of hazardous substances to which we are exposed. But it will do little to correct the problem at its *source*, which is that petrochemical production is governed not so much by society's needs for the products as by the industry's need to make a profit by manufacturing a growing variety of substances in the largest possible amounts.

On the other hand, if sufficient attention is given to the act's broad mandate to compare the benefits and risks of a petrochemical product with those of available substitutes—including natural ones—and to consider the economic and social consequences of choosing between them, the hearings may have an entirely different outcome. Industry representatives are then likely to reiterate their belief in the benefits of their products, relative to the risks: that agricultural chemicals produce plentiful food at reasonable prices; that synthetic textiles produce affordable clothes which are easy to maintain; that plastic auto parts are lighter, stronger and cheaper than metal ones; that the construction of a plant to manufacture a new plastic substitute for leather will create new jobs. In rebuttal, their opponents may well reply that recent studies show that some commercial farmers do about as well with organic methods as their conventional neighbors who use chemicals, that cotton fabrics are much less energy intensive than synthetic ones (which are, after all, made from petroleum and natural gas); that the economic gains afforded by cheaper auto parts are offset by the economic drain imposed on a capital-short economy by the capital-intensive petrochemical industry; that because leather production is much more labor intensive than the petrochemical industry, the displacement of leather by a new plastic will reduce, rather than increase, the total number of available jobs.

Guided by this broader mandate, the Toxic Substances Control Act can do much more than regulate the use of hazardous materials. It can, in fact, open what the industry may regard as a political Pandora's box, providing an arena in which the public will be able to intervene effectively in a process that is normally the exclusive domain of industrial managers: decisions about what to produce and how to produce it.

This issue has already been raised in connection with nuclear power. Concern over its hazards had led to a growing demand for phasing out the industry; its social burden has generated a demand for social governance of the industry. Many Americans already hold strong opinions about one major petrochemical product, plastics. A recent public-opinion poll conducted for the Society of the Plastics Industry showed that, in comparison with alternative goods, many public officials regard plastics as the most hazardous alternative, and the product which they would be most willing to do without.

Based upon widespread concern over health and environmental problems, the public appears to be ready—as it is in the case of nuclear power—to determine what balance between the hazards and benefits of the petrochemical industry is acceptable. Not only in terms of health and the environment, but also in terms of the economy and social welfare; and not only of the petrochemical products themselves, but also in comparison with the alternative natural products which have been the chief victims of the industry's compulsive invasion of the economy.

In this way the Toxic Substances Control Act can facilitate a discussion which will enable the people of the United States to decide how to respond to the sweeping challenge of the petrochemical industry. Shall we establish another bureaucracy that attempts, in long and costly deliberations, to alleviate the hazards to health and the environment that are so closely associated with the petrochemical industry? Or shall we establish some means of social governance over the forces that compel the industry to bestow upon us, along with its benefits, the biological dangers that are inherent to an enterprise bent on confronting living things with substances that they, in their evolutionary wisdom, have long since rejected?

Alaska and the Pipeline: A Picture Essay

The Alaska pipeline, running for 800 miles from Prudhoe Bay to Valdez, opened for business this year. In keeping with the sheer magnificence of its host state, the pipeline has been one of the most ambitious projects ever completed. And in spite of countless environmental, technical, political and social hurdles, we are beginning now to enjoy the first fruits of oil from the North Slope.

Because the oil is in one of the most remote regions of the world, the cost of building the pipeline was enormous. The North Slope of Alaska is bordered by the Arctic Ocean which is frozen for at least nine months of the year. The Brooks Range, to the south of Prudhoe Bay, forms a 100-mile wide barrier for workers. Winter temperatures average about 20 degrees below zero. The wind chill factor can effectively lower that average to the point that ice masks can form from the breath of each worker, creating surrealistic shapes.

The fragile environment of Alaska was an important factor in considering and planning how the pipeline was to be built. The ground is in a state of permafrost covered by tundra. The caribou follow established routes in their migratory patterns across the state—routes that could not be blocked by the 48-inch pipes carrying oil to the south.

To accomodate all the forms of life that Alaska supports, the pipeline was designed to carry oil above and below ground. One special consideration was the state of permafrost that exists in the northern reaches of the pipeline. To bury the line would jeopardize the effective passage of oil since freezes and thaws could cause dangerous stress on the pipes possibly leading to breaks and severe oilspills.

Alaska is a rich and varied land of great extremes. It has 33,904 miles of saltwater coastline; 3 million lakes measuring over 20 acres; 10,000 rivers and streams; four time zones; and the highest mountain peak in North America—Mt. McKinley, which is 20,320 feet high. It has a rich and varied wildlife, including wolves, the Kodiak Bear, caribou, polar bears, moose, elks, seals and all kinds of birdlife. The next pages show the spectacular beauty of the state and some of the work that led to the construction of the Alaska Pipeline.

Chukchi Sea

Cape
Beaut

Point Hope ●

ARCTIC CIRCLE

Bering Strait

Kotzebu

SEWARD PENINSUL

● Nome

Bering Sea

The pipeline route travels across some of the most rugged terrain on earth. Note especially the vast Brooks Range just south of

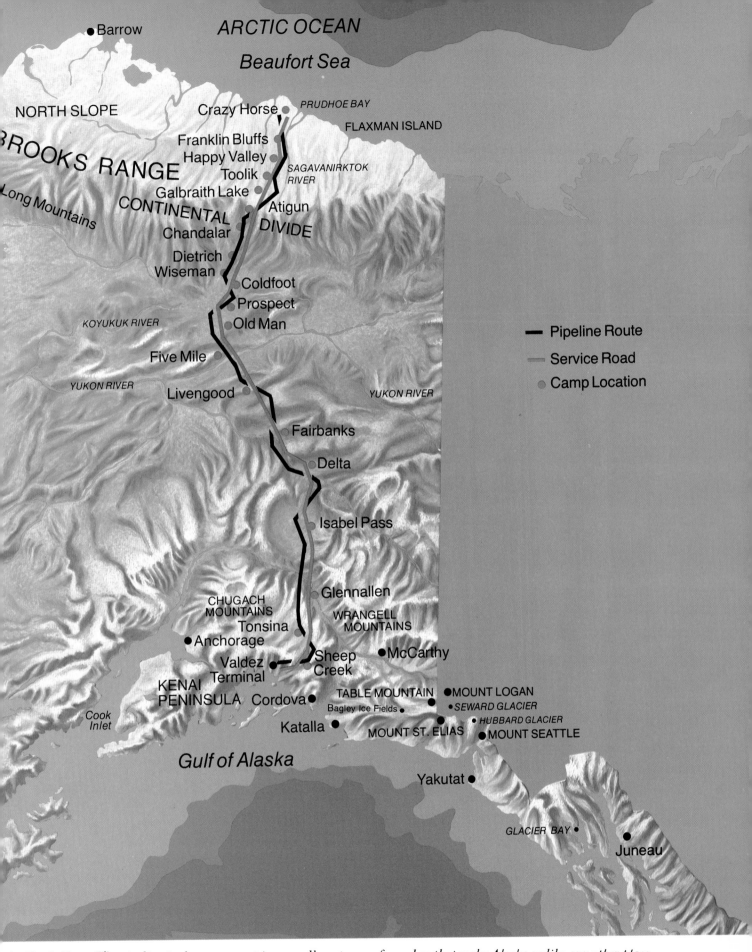

ARCTIC OCEAN

Beaufort Sea

Barrow

NORTH SLOPE

Crazy Horse

PRUDHOE BAY

FLAXMAN ISLAND

BROOKS RANGE

Franklin Bluffs

Happy Valley

SAGAVANIRKTOK RIVER

Toolik

Galbraith Lake

Long Mountains

CONTINENTAL

Atigun

DIVIDE

Chandalar

Dietrich

Wiseman

Coldfoot

Prospect

KOYUKUK RIVER

Old Man

Five Mile

YUKON RIVER

Livengood

YUKON RIVER

Fairbanks

Delta

Isabel Pass

— Pipeline Route

— Service Road

● Camp Location

Glennallen

CHUGACH MOUNTAINS

WRANGELL MOUNTAINS

Tonsina

Anchorage

McCarthy

Valdez

Terminal

Sheep Creek

KENAI PENINSULA

Cordova

TABLE MOUNTAIN

MOUNT LOGAN

Bagley Ice Fields

SEWARD GLACIER

Cook Inlet

HUBBARD GLACIER

Katalla

MOUNT ST. ELIAS

MOUNT SEATTLE

Gulf of Alaska

Yakutat

GLACIER BAY

Juneau

the North Slope. The pipeline is the most recent in an endless stream of wonders that make Alaska unlike any other place.

Top left. *Wearing a new coat of high-visibility orange, the U. S. Coast Guard Icebreaker* Glacier *takes scientists deep into the Beaufort Sea pack ice.* Top right. *An unnamed glacier containing ogives, or seasonal bands of airborne dust, flows over Thompson Ridge.* Lower right. *Ernest de Koven Leffingwell, the first person to accurately chart the North Arctic coast of Alaska, lived in this small cabin built from the timbers of a wrecked sealing schooner. Recently the tiny house was named to the National Register of Historic Places. C.G. Mull affixes the plaque to the side of the cabin.* Lower Left. *A lone polar bear bounds across the Beaufort Sea Pack ice. Unlike most arctic seas, the Beaufort has been described as a "cold, biological desert." It sustains relatively small proportions of marine and land life.*

From 1906 to 1914, Ernest de Koven Leffingwell's spartan life was based within this modest cabin.

An Exxon geologist, John Leftwich, samples rocks in southern Alaska's St. Elias Range as part of a continuing study of the state's geological past.

Left. *An aerial view of the pipeline as it stretches off to the distance.* Top left. *Prudhoe Bay may be the largest oil field in the world. It is surely one of the most remote. Here a scientist scans a white skyline relieved by a line of oil wells.* Top right. *Five-ton sections of 48-inch pipe were stored near Valdez, Alaska for construction of the pipeline. A spectacularly beautiful inlet, Valdez has mild winters and stable bedrock for building facilities. The deep natural harbor made this town of 600 people the terminal point of the Alaska pipeline and the* shipping center to the southern marketplace. Lower right. *Overlooking Port Valdez, the terminal has two tank farms to contain Prodhoe Bay's crude oil. There are 18 tanks capable of holding 500,000 barrels of oil each.* Lower left. *Since the possibility of oil spills from the pipeline is present in spite of hundreds of safety measures, scientists tested the effect of crude oil on vegetation. Here are test plots where 2½ to 5 gallons of oil were applied per square meter to determine the damage on fragile grasses.*

Mining the Wealth
of the Ocean

by William Wertenbaker

The floor of the sea is a land no man can ever walk, a spectacular landscape miles beneath the ocean's surface, only recently explored and, as late as 25 years ago, virtually uncharted. Much of the surface of the moon is better known than the surface of the earth beneath the sea. But the floor of the sea has become the subject of one of the bitterest—and, some observers say, most hopeless—disputes in international relations today. On the bottom of the ocean are mineral deposits large enough to supply all mankind for years, even centuries, to come. There is enough copper to last the entire world for 6,000 years, compared with 40 years' reserves on land; enough nickel to last 150,000 years, versus 100; aluminum for 20,000 years, versus 100.

With deposits on land dwindling and becoming harder to find and more expensive to mine, some of the largest metals and energy companies in the world have formed partnerships to mine the sea for metals that today cost the United States some $2 billion yearly to import. Some companies have already mined the floor of the sea successfully, but only in tests. One company reports two areas which alone hold more nickel than all the known reserves on dry land. Another company tried to file a claim with the United States Government for the rights to a "mother lode" on the Pacific Ocean floor.

Altogether, the companies have spent more than $100 million on exploration and testing. Now they want to begin ocean mining commercially.

It's a hunt for sunken treasure on a corporate, national and global scale. But what are the rights of the hunters? Who is to control the sea floor— an area larger than all the continents together—and who is to profit from it? These questions have embroiled more than half a dozen meetings of the United Nations Law of the Sea Conference.

More than half of the world lies two miles deep beneath the sea. For a generation, in increasing numbers, men in small ships have prowled the oceans, profiling, photographing and sampling the floor of the sea, drawing the shape of a landscape no one had ever seen, finding new mysteries. They discovered the forces that shape the continents and the causes of mountain ranges, eruptions and earthquakes, all of which had baffled geologists for a century. Beneath the sea, clefts, cliffs, plains and mountains lie in pristine grandeur. Plains flatter than the eye can measure stretch over the curve of the earth. Low hills spring from the plain and cluster on higher ground. Seamounts rise out of the landscape, their flanks sheer and unencumbered by foothills. Gaping chasms drop as deep again as the sea floor itself. Massed peaks rise rank on rank into a chain of mountains greater than any above the sea.

On the ocean floor it is dark, silent and almost still. Placid currents whiffle by, leaving ripples in the sediments. There are fish and shrimp in the water; anemones, starfish, sea urchins and odd animals called sea cucumbers on the bottom, clams and a myriad of tiny worms in the sediment itself. More happens in the deep sea than anyone ever suspected. Time-lapse photography in a part of the Pacific staked out for deep-sea mining has surprised biologists: The bottom visible to the camera changed completely in six months; rough places became smooth, and smooth ones rough. Animals stalked by, leaving only their tracks. Something that for three months had looked like a rock got up and moved about a foot, then settled down again and looked like a rock for three more months. Another rocklike thing sprouted an arm and waved it about for 12 hours, then remained motionless for the rest of the six months.

Life proceeds without haste in the deep. A tiny clam may not become sexually mature and reproduce until it is 60 years old. Decay (including the biodegration of pollutants) is nearly imperceptible: Year-old sandwiches found in a lunchbox in a sunken research submarine were entirely fresh—though soggy. The strange life of the sea floor could be a unique resource. Biologists have already produced drugs from similar creatures of shallower waters: a potent, safe insecticide and several substances that inhibit cancers in laboratory experiments. The animals of the deep abyss have yet to be examined.

Among the mountains of the sea, and spread across the hills and plains, vast deposits of minerals have been found and sampled. According to one authority, the deep-sea sediment called red clay contains a quantity of copper beyond comprehension—a trillion tons or so—which if it could be dredged up profitably might supply the world for thousands of years. The floor of the sea, moreover, is the ultimate source of many, if not most, of the mineral deposits now mined on land. One of the most remarkable discoveries of the last decade is the way in which the floor of the sea has been—and is still—transformed into dry land during the creation of a mountain range. With this new understanding geologists expect to find new mines ashore.

MANGANESE NODULES

But the immediate commercial focus is on manganese nodules, strewn across the sea floor as thick as windfall apples. Manganese nodules are round, or potato-shaped, or like bunches of grapes, or like hamburgers—or shaped liked nothing in particular. The nodules contain 40 different metals, including copper, aluminum, nickel, iron and cobalt—as well as manganese. Copper is used for electrical wiring, tubing, roofing, paint pigments; nickel for coinage, safes, armor plate, in paints and varnishes. Iron is used to make steel, cars, machine parts, stoves, pipes, radiators. Cobalt, now being used to treat cancer, is a component of stainless steel and of alloys used to make jet engines and cutting tools. Manganese is used to harden steel and in making glass, paint, fertilizer.

There are an estimated one and a half *trillion* tons of manganese nodules on the ocean floor. Their numbers seem endless. They are surprisingly light, porous and crumbly. Mostly they are black, but sometimes brown. Most are half an inch to six inches across. The largest nodule ever found weighed almost 1,900 pounds; it came to the surface tangled in a telegraph cable and was thrown back. Manganese nodules were first seen by the British ship Challenger, which made the first long oceanographic voyage a hundred years ago. There were nodules in most of the samples the ship dredged from the sea floor throughout the world; at first the scientists aboard thought they were fragments of meteorites. Not until after World War II, when the first deep-sea cameras were put into use, did anyone see the overwhelming numbers of manganese nodules carpeting the ocean floor, often packed so thick there is no space between them. They lie in fields that cover millions of square miles in every ocean. There is even a little field of nodules in Green Bay, Lake Michigan.

Manganese nodules, unlike most ores, contain little but metal. Although they cover more of the earth's surface than almost any other salable resource and have been known to science for a century, scientists still don't know how they are formed. They are not fragments of meteorites, though they sometimes look like cinders. Microscopic creatures may have a significant role in their creation. Like pearls or laboratory crystals on a string, they grow around a seed, a scrap of

Deep-Sea Mining Techniques: Two ships tow a 16,000-yard rope carrying a series of buckets that scrape up nodules. Deepsea Miner II uses compressed air to lift nodules vacuumed off the ocean floor, three miles below. Another method employs a dredge to scoop and either compressed air or a hydraulic pump to raise the nodules.

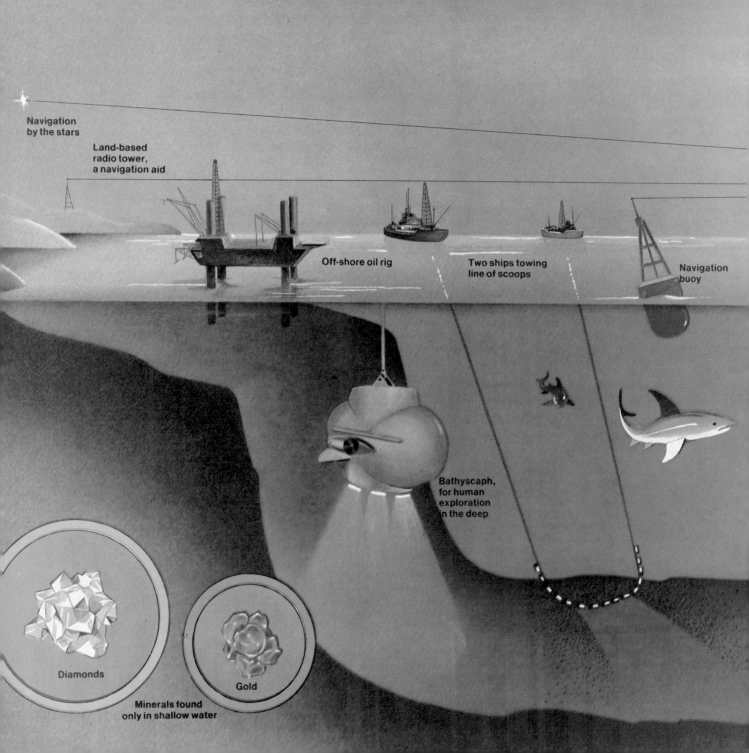

Navigation
by the stars

Land-based
radio tower,
a navigation aid

Off-shore oil rig

Two ships towing
line of scoops

Navigation
buoy

Bathyscaph,
for human
exploration
in the deep

Diamonds

Gold

Minerals found
only in shallow water

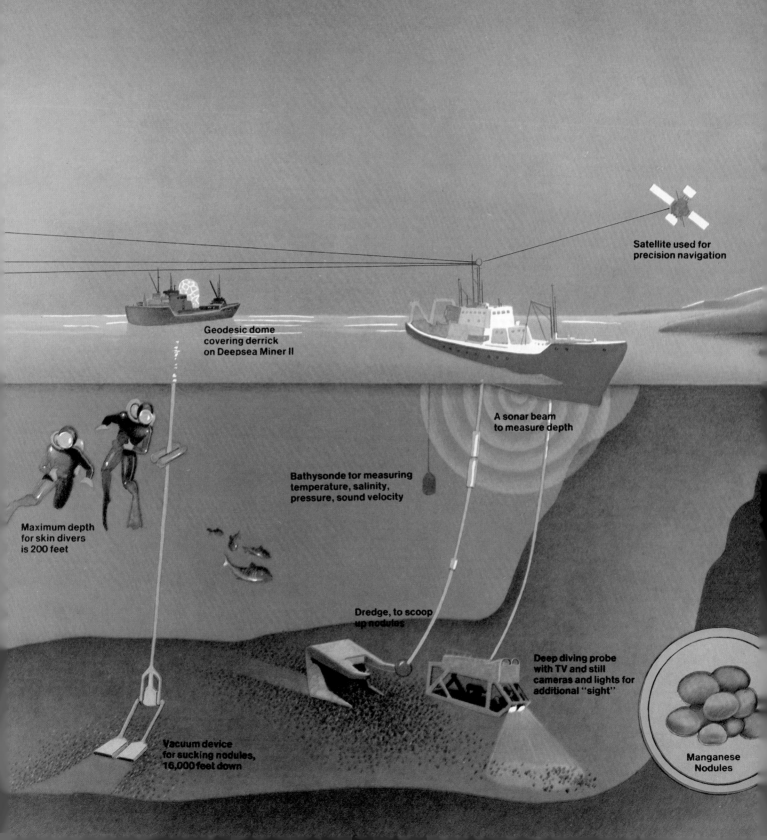

Satellite used for
precision navigation

Geodesic dome
covering derrick
on Deepsea Miner II

A sonar beam
to measure depth

Bathysonde for measuring
temperature, salinity,
pressure, sound velocity

Maximum depth
for skin divers
is 200 feet

Dredge, to scoop
up nodules

Deep diving probe
with TV and still
cameras and lights for
additional "sight"

Vacuum device
for sucking nodules,
16,000 feet down

Manganese
Nodules

A dredge basket filled with manganese nodules is unloaded after a television survey of the ocean floor.

The television camera transmits this image of an actual nodule deposit 18,000 feet beneath the ocean surface.

bottom sediment, a shark's tooth, a whale's earbone, a flake of rock, a piece of another manganese nodule. When a nodule is sawed in half, it reveals a "seed," showing growth layers around it like tree rings. The nodules grow very slowly. A hundred-pounder pulled up by a ship of the Scripps Institute of Oceanography in San Diego was found by laboratory analysis to be about 16 million years old and to have grown a millimeter per 100,000 years. Nonetheless, nodules en masse have been estimated to be growing faster than mankind can use them—or, by a less enthusiastic calculation, at the more modest rate of 16 million tons per year. About the only areas in the ocean where nodules are not found are those where sediment accumulates faster than nodules can grow, burying the

incipient nodules before they are half begun.

The metals found in nodules are the same ones that are found dissolved in seawater and distributed through the sediments on which nodules lie. Whether they come from the seawater or the sediments or from both, however, is still a matter of speculation. One theory is that minute, snail-like insects called forams, which sometimes coat nodules (like barnacles), act as a sort of catalyst in the growth of nodules. Another theory is that bacteria are the active agents, and yet another that the amount of rock fragments spread about by undersea eruptions determines where there shall be nodules, and how many. Whatever the cause, nodules vary from one place to another. Some regions have few; others, no room for more.

Some have small ones; others, large. In some places, nodules are richer in certain metals than in other places. Nodules from the Pacific are softer than those of the Atlantic.

DEEP-SEA MINING

While gold and diamonds have been mined from shallow water for many years, manganese nodules, locked in a remote, forbidding environment, were only of academic interest until 1959. In that year, John Mero, then a young graduate student in mining engineering at the University of California, published the first effective argument for deep-sea mining. Mero calculated where nodules were most plentiful, how many there were, what they were worth, how they could be mined and at what cost. A stocky, brown-haired man now in his mid-40's, Mero has been writing and lecturing on the potentials of nodule mining ever since. He is now developing a promising method of ocean mining. People familiar with the field have called him everything from a far-sighted pioneer to an opportunist, but it is probably more than coincidence that the ocean-mining projects now under way date back to within a few years of his original report.

Far from being a last resort, mining the floor of the sea has certain advantages over mining ashore. A deposit of manganese nodules can be found without drilling or blasting; and every nodule to be mined during a 20-year-long project can be photographed and counted with still and television cameras before a single piece of mining machinery is ordered. No shafts have to be bored or mountains bulldozed to reach the ore, no power plant constructed to process ore on the spot, no railroad built to carry it to the ocean for shipment abroad, no towns erected to house miners. The cost of a deep-sea mining operation could be as much as $500 million, but in recent years new mines in remote places on land have cost nearly twice that. Mero points out that there are no cartels at sea, like OPEC, to manipulate prices; no hostile or unstable governments to seize assets, and no unfavorable balance-of-payments problems. He claims that deep-sea mining would keep metal prices stable for years. Four of the metals found in nodules — copper, nickel, manganese and cobalt — are considered commercially interesting today, and

the United States imports all of them at a cost of about $2 billion a year. About half of that amount could come from ocean mining by 1985, according to various estimates.

MINING BY CONSORTIUM

The corporations involved in deep-sea mining are American, British, French, Belgian, German, Dutch, Australian and Japanese. There are four mining ventures active in this country, each backed by a consortium of companies, as well as projects in France, Germany and Japan. The Virginia-based Deepsea Ventures Inc. is backed by a consortium of United States Steel, Sun Company (formerly Sun Oil) and Union Miniere of Belgium. The New York-based Kennecott Copper Corporation is the managing partner in a consortium with British Petroleum, Rio Tinto-Zinc Corporation and Consolidated Gold Fields Ltd., all of Britain, Noranda Mines Ltd. of Canada, and Mitsubishi Corporation of Japan. International Nickel Company, Metall-gesellschaft A.G. and other German companies, Sumitomo and a subconsortium of Japanese Companies, and Sedco Inc. are partners in Ocean Management Inc. of Bellevue, Wash. Lockheed, Amoco Minerals and a subsidiary of Shell Petroleum have agreed to consort and are looking for another partner. International Nickel, United States Steel and 14 other companies are also backing the development of a mining system by John Mero's Ocean Resources Inc., based in San Diego. The Summa Corporation, which claimed it was mining manganese nodules with the ship Glomar Explorer, but which was really trying to salvage a Russian submarine for the C.I.A., has since disbanded its mining division, according to reports.

The financial risks in ocean mining are substantial, and of the dozens of companies involved, none has yet elected to undertake them alone. While failures are to be expected in any new venture, complicated machinery tends to break down at sea. One mining expert compares his company's process to picking something off the sidewalk by dangling a long piece of spaghetti from the top of the World Trade Center. The simplest of the systems being developed is John Mero's continuous line bucket, a series of one-ton hoppers on 16,000 yards of four-inch-thick rope. The line, which hangs between two ships, is towed slowly, so that each bucket drags along the bottom and scoops up the nodules.

The four consortia are developing more complex and expensive—but perhaps also more effective—systems than the line and bucket. Suction dredges winnow nodules from the seabed and then suck them up a pipeline by means of hydraulic pumps or compressed air. The consortia have built dozens of components and cautiously tested them piece by piece in different shapes, sizes and combinations; they have tested small-scale mining devices at sea and small-scale plants for processing nodules into metal ashore. They have even constructed simulated sea floors in which to make more tests. In 1970, Deepsea Ventures spent several

months mining nodules from the Blake Bahama Plateau, a relatively shallow part of the Atlantic only a few hundred miles from the company's headquarters in Virginia. Though only a test, it was the first time nodules had actually been mined. Kennecott, which is building the largest mining system (to handle 15,000 tons of nodules a day, whereas Deepsea is aiming for 5,000), has operated equipment in much deeper water—but without trying to bring up any nodules.

All the consortia have scheduled what they consider major tests during the next year or so. One ship, however, is about to embark on a shakedown cruise in the Pacific that should yield 15,000 tons of nodules for Deepsea Ventures Inc. An ore carrier 560 feet long, converted for the job and renamed R.V. (for

The Deepsea Miner II provides a platform to conduct deep ocean mining tests.

research vessel) Deepsea Miner II, it can take 1,000 tons of nodules a day off the bottom, a fairly close approximation of commercial mining, though still at only one-fifth its capacity. It's a queer-looking vessel, with a large geodesic dome sprouting out of its middle. Inside is a derrick, and a well in the bottom of the ship through which equipment can be lowered to the bottom of the sea. Oil drillers call such a well a moonpool—after its nighttime reflections—but Deepsea has roofed out the moonlight. The derrick is gimbaled, or suspended, so the rocking of the ship doesn't break the pipe that brings the nodules up from the bottom.

THE GOLDEN HORN

The best nodule fields in the world, both for quantity of nodules and quantity of metals in the nodules, lie along a strip of the Pacific Ocean floor 2,500 miles long and 800 miles wide. The strip starts at a chain of seamounts about 600 miles west of the coast of Mexico and runs to the Line Islands, around 600 miles southwest of Hawaii. Identified in 1972 by David and Barbara Horn and Maryland Delach of Columbia University's Lamont-Doherty Geological Observatory, the strip is called the golden horn. There the mining consortia have been running surveys with high-precision sonar and still and television cameras, looking for the richest pockets and then sampling and mapping them in minute detail. John Mero originally estimated an ocean-floor nodule mine could be operated for 20 years on 20,000 square miles of sea floor. There are two *million* square miles in the golden horn.

In 1972, Deepsea Ventures, always a little more flamboyant than its competition, addressed to Secretary Kissinger and the State Department what it called a claim to the mining rights of just over 20,000 square miles of the golden horn. (The State Department replied that it does not register or recognize claims.) In late 1977, Deepsea Miner II will arrive there. Its crew will lower a dredge through the moonpool by attaching some 450 sections of pipe, one by one, until the dredge hits the ocean floor—16,000 feet below. The process will take five days. Then the ship will tow the dredge at the speed of two to four miles an hour, scraping and sucking up nodules, across a 6-by-19-mile

180

section of the low, rolling country that Deepsea has mapped by computer. Sonar transmitters will be set on the bottom along the way to help navigation, and there will be two television cameras on the dredge itself looking ahead at the nodules and any obstructions that may lie in the way. Bow and stern thrusters on the ship will move it crabwise if necessary.

Even so, the miners don't expect to "lawnmower" the bottom. (This is a lawnmower that bags its cuttings.) The best they can do in three miles of water is to cut a swath that looks as if it had been mowed by someone who had drunk six martinis. They don't really know the environmental impact of what they are doing. The animals in the path of the dredge will probably be killed, others may be suffocated by stirred-up sediment. Scientists say that although the effects of the sea floor on the rest of the world are unknown, it is unreasonable to expect half of the world not to have a significant effect on the other half, whether or not we are clever enough to see it. A few years ago, two distinguished scientific journals speculated that spores, bacteria and other small and noxious creatures that had lain dormant on the ocean bottom for centuries would "bloom" in the surface water after being dumped off a mining ship. No "blooms" have actually been observed, however. According to Anthony F. Amos and his associates, oceanographers of the University of Texas who saw the mining in the Blake Bahama Plateau, the concentration of sediment dumped in the ocean was .05 percent after eight minutes.

ECONOMICS OF MINING

The profits of nodule mining are either excellent, marginal or imaginary, depending on whom you talk to. "The economics are very, very favorable," said John E. Flipse, who resigned as president of Deepsea Ventures. The economics will turn on nickel, a metal that is both expensive—more than $2 a pound today—and so plentiful in nodules that two companies could be mining 40 percent of what the nation uses by 1985, according to studies. Kennecott anticipates producing 80 million to 90 million pounds of nickel a year and should earn $100 a ton from the three million tons of nodules a year it proposes to mine. Kennecott claims it will cost

$5 to $10 a ton to mine nodules and $10 to $15 to refine them. Independent analysts double those figures, add transportation and investment costs and still find a profit. Richard Tinsley of the Continental Bank in Chicago, who has made a specialty of ocean mining, estimates it will cost $60 to $70 to mine a ton of nodules and process it into metals worth $90 to $100.

The Ocean Mining Administration of the United States Department of the Interior is more pessimistic. It has concluded that nodule mining could turn an annual profit of 12 to 22 percent, which "may be considered by some barely adequate to justify the ... risks," Some mining specialists go further and say nodules will not be mined for many, many years; according to one, "What they are doing now is a tax write-off. They're not going to compete against their own [on-shore] mines. They'll say ocean mining isn't economic and charge what they are doing to the taxpayer." However, the investments—$50 million for Deepsea Ventures alone, according to Flipse—may already be high for this theory; and the mining consortia already can blame the U.N. for any delays or changes of heart—and they have.

A television camera and tripod are launched over the side of a research vessel. A manganese nodule sample basket at the foot is used to recover small quantities of nodules for analysis.

THE LAW OF THE SEA

On Nov. 1, 1967, the delegate from Malta, Dr. Arvid Pardo, spoke to the United Nations General Assembly on the global problems of the use and misuse of the oceans. Detailing the vast wealth blanketing the seabed, he reminded nations of the loopholes in existing international law on the sea. That law, in the Geneva Convention of 1958, is a model of evasion couched in high-sounding language. It gives coastal nations sovereignty over the ocean floor out to a depth of 200 meters, the edge of the continental shelf—"or beyond that limit, to where the depth of the superjacent waters admits of the exploitation of the natural resources of the said area." In translation, the law says whatever is on the floor of the sea belongs to whoever can get it. "That's almost free license," Pardo remarked later. Under the law, the United States could claim ownership of all the Pacific from the golden horn area to Hawaii and the mainland—or almost anywhere

else. "We could see a scramble for land like the one in Africa in the last century," said Pardo. Raising that specter before the small nations, Pardo called on the General Assembly to pass a resolution saying that the resources of the sea floor were a "common heritage of mankind." In 1969, the U.N. passed such a resolution, also making the floor of the sea off-limits to mining until further notice. But the resolution is not law, and all the international conferences held since 1973 have failed to agree on an international regime over the sea floor.

Frustrated by many countries' suspicions that poor nations would be euchred out of the sea's great wealth by the rich, many diplomats entered the latest Law of the Sea negotiating session in May 1977 with little optimism. The 114 third-world nations, represented by the so-called "Group of 77," would have the "common heritage of mankind" husbanded by an international body generally referred to as the Authority, under which there would be another body, known as the Enterprise, doing all the mining. Each United Nations country—maritime or landlocked—would have one vote in the policies of the Authority, and no commercial enterprise would be allowed to mine the floor of the sea. But industrial nations like the United States have insisted that their companies be allowed to mine, too, and they have proposed various compromises which, according to supporters of the Enterprise, would hamstring its operations.

"Rarely has any generation had so clear a choice to make between order and anarchy," said Elliot Richardson, the chief American delegate to the United Nations Law of the Sea Conference, as this year's negotiating session opened. As the session neared its end, considerable progress had been made, though major issues were still unresolved. But participants were privately astonished that even with but a few days still remaining there seemed an equal chance either of the session's resolving those issues or of its ending with a major setback. As for the mining companies—who call progress at the U.N. slower than the growth of manganese nodules—some claimed to have intentionally delayed their projects because of the unclear status of their mining rights. Three separate bills have been introduced in Congress, however, to resolve the legal uncertainties caused by the absence of a Law of the Sea. And observers confidently expect this Congress to pass one of them. "With the law straightened out," says the head of one mining group, "we could be doing real mining in a couple of years."

Too Hot to Handle

by Richard Severo

A night during the week of Christmas 1969. Yuletide cheer in the home of a worker employed in the nearby nuclear recycling plant at West Valley, N. Y. Suddenly, men appear at the door. They enter and take away the worker's boots, the living room rug, a bedsheet, a baby blanket and other household items. The confiscated property is rushed off, some of it to be decontaminated, the rest of it to be buried in the earth. The plant's safety precautions had slipped. The worker has been tracking dangerous radioactivity over the countryside and through his home.

"We are," said Gov. Nelson A. Rockefeller, "launching a unique operation here today which I regard with pride as a symbol of imagination and foresight...." The day was June 13, 1963, the place was a sparsely settled area near the hamlet of West Valley, N.Y., and the speech showed Rockefeller at his decisive and optimistic best. He was officiating at the groundbreaking of the world's first commercial nuclear-waste plant—a $32.5-million facility that would take the spent fuel of the atomic-power industry and reprocess it for renewed use. The plant, to be operated by a newly created private company, Nuclear Fuel Services, would sell its recycled fuel to civilian atomic-power plants around the country, and the operation, said Rockefeller, would "make a major contribution toward transforming the economy...of the entire state."

Governor Nelson A. Rockefeller

183

Today, 14 years later, the plant, reduced from full-scale operation to a skeleton crew of 50, sits silently in the undulating landscape 30 miles southeast of Buffalo, a technological and economic disaster, and Rockefeller is a disappointed man. Looking back at the misadventure that may end up costing the taxpayers of New York half a billion dollars, or a billion dollars, or even more, depending on whom you talk to, he concedes: "Obviously, this is not the answer, and there's no question that we've got a new problem: what do you do with the stuff?" The "stuff" is the 600,000 gallons of highly radioactive liquid wastes that are today contained in a tank buried in the ground nearby, and the 2 million cubic feet of buried radioactive trash into and out of which water has leaked, spreading radioactivity into Cattaraugus Creek, which flows into Lake Erie, from which the city of Buffalo obtains its drinking water.

It would be easy to blame the fiasco at West Valley on the impetuosity of a Governor proud of being "action-oriented," unwavering in his belief in the credo of technology—that society must take risks if it is to progress and accept losses if it is to learn. But the story that emerges from an extensive investigation of this nuclear enterprise is one in which the blame is shared by many, many others—by everybody and by nobody.

UNCERTAIN BEGINNINGS

It is the story of technocrats who assured and reassured the public that nuclear recycling was safe and that a thoughtfully engineered fail-safe system would minimize the hazards of any accidents that might possibly occur—without

WHAT ARE ISOTOPES

ISOTOPES ARE ATOMS OF AN ELEMENT
DISTINGUISHABLE BY THEIR WEIGHT

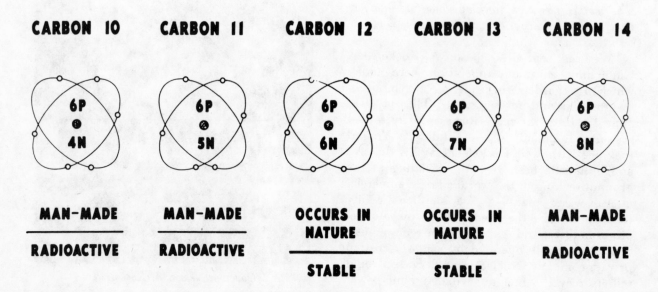

CARBON 10	CARBON 11	CARBON 12	CARBON 13	CARBON 14
6P 4N	6P 5N	6P 6N	6P 7N	6P 8N
MAN-MADE	MAN-MADE	OCCURS IN NATURE	OCCURS IN NATURE	MAN-MADE
RADIOACTIVE	RADIOACTIVE	STABLE	STABLE	RADIOACTIVE

making it clear that their assurances were based on extrapolations from premises rooted in probabilities and anchored in uncertainty. It is the story of company officials who repeated such assurances even after scores of incidents—known only inside the company and to a few Government inspectors—had made it clear that leakage of radioactivity within the plant was reaching dangerous levels.

From former employees interviewed in the course of the investigation, and from the files of the Atomic Energy Commission, come accounts that give an insight into some of the things that went wrong:

☐ A malfunctioning vent spewed radioactivity into a workers' lunchroom.

☐ A worker with contaminated hair was advised to have a haircut without telling his barber about the "problem."

☐ Radioactive tools were "borrowed" for use outside the plant; a laundry room was evacuated because of radioactivity; radioactive water went by mistake through regular drainage pipes.

☐ Workers showed up drunk for the night shift; a security guard, in a complex containing radioactive waste, fired wildly at a fox.

Rumors of these and similar "incidents" to be detailed in this article filtered through the farm country in which the plant is set. But, because of the isolation of rural living, these rumors remained indefinite and unconfirmed, and A.E.C. officials who investigated the incidents notified the company and then apparently filed the reports away. What little knowledge there was in the West Valley area was remarkably slow to spread. Because the story is also one of working people who wanted jobs, public officials who wanted taxes and businessmen who wanted economic benefits—all of them wanting to believe that the plant was safe and that its potential was great.

CITY OF OZ

Finally, it is the story of a dream gone sour—the dream born in those days after World War II when Americans saw a City of Oz in their future, a place where the nation's energy needs would be met by the wizardry of atomic energy. And there was money to be made. The nuclear

fuel is uranium. Nuclear fuel not totally used up the first time around could be reprocessed, recovering not only unused uranium but plutonium, which is created from uranium in the reactor process. (The Government already had been doing some reprocessing of the uranium it uses for making nuclear weapons). The extracted uranium and plutonium would be used as a new fuel. Our supplies of uranium—our source of nuclear energy—would be stretched out.

The country now has 64 operable nuclear power plants, according to the Atomic Industrial Forum, plus 72 under construction, 19 on which limited work has been done, 64 ordered, and seven more suggested by letters of intent. By the year 2000, if present plans hold, there could be between 200 and 300 such facilities in operation. Now, however, there is no certainty of agreement on what to do with the perilous wastes they produce. And the expectation that the wastes can be reprocessed into new sources of atomic energy has been dealt a heavy blow at West Valley, heavier than government and industry have cared to admit. In the light of this investigation of what was to have been the pioneer project of a whole new commercial industry, the technology surrounding the reprocessing and storing of nuclear wastes stands exposed and awkward, much the way the Wizard of Oz did when the curtain was pulled aside to reveal not a real wizard but a fallible if well-intentioned man.

THE PEACEFUL ATOM

It all seemed promising enough in 1963. Nuclear Fuel Services, formed as a subsidiary of W. R. Grace's Davison Chemical Company, built its plant on a part of 3,331 acres of land leased from New York State; the land had been purchased from farmers under the right of eminent domain as part of Governor Rockefeller's plan to place the state in the vanguard of peaceful atomic development.

The contracts, it was true, seemed curious even then, since they shifted virtually all of the risk of a profit-making venture away from the company and onto the shoulders of the state's taxpayers. Section 26.01 of the lease said that, whatever happened, N.F.S. would not be required "to remove radioactive contamination from the leased premises." And section 3.04 of the waste-storage agreement, signed with the state's Atomic Research and Development Authority, spelled out the taxpayers' liability even further: "Upon any cancellation or termination of the leases . . . the Authority will assume full responsibility for perpetual operation, surveillance, maintenance, replacement and insurance of the then high level [intensely radioactive] storage facilities." The contract said this was to protect the "health and safety of the public."

N.F.S. was required to set aside some money to be placed in escrow as a "reserve for perpetual care and management of nuclear wastes." The funereal-sounding "perpetual care" referred to the fact that some nuclear wastes remain dangerous for hundreds or thousands of years. The escrow account now holds less than $4 million—a modest amount, in view of the estimated $500-million to $1-billion it would take to decommission the plant and clean up the mess. Representative Richard L. Ottinger of Westchester County, a critic of nuclear reprocessing in general and of N.F.S. in particular, says the state "practically gave the bank away" in signing those contracts. Rockefeller says he wanted to provide jobs for a state that needed them, and the contracts "may have been the only way we could get the operator [N.F.S.] to come in."

The plant opened in 1966. In its six years of operation, it reprocessed about 625 metric tons of nuclear fuel. Some of it came from commercial atomic power plants in Michigan, Minnesota and Puerto Rico. But since initially there weren't enough civilian wastes to keep the plant going, a deal was worked out to take wastes processed by the Government-owned reactor in Hanford, Wash. More than 60 percent of all the wastes processed by N.F.S. during its lifetime came from there. Some of the recovered plutonium went back to Hanford, apparently for the manufacture of bombs and other military purposes—a detail that would come as a surprise to those New Yorkers who took pride in what they saw as the state's pioneering role in the peaceful use of the atom.

PROFITS ON THE HORIZON

In 1969, Davison Chemical Company sold N.F.S. to Getty Oil. The plant had been a

perennial money loser, but Getty, according to N.F.S. general manager William Oldham, saw the investment as "patient capital": Sooner or later, the nation's growing demand for energy, coupled with its dwindling oil supplies, would force Americans to turn more and more to nuclear energy. And that would mean more and more nuclear wastes to process. Profits, they thought, were on the horizon.

As for the site itself, company representatives said it lent itself well to reprocessing, storage of radioactive liquid wastes and burial of radioactive trash (such as the uniforms of atomic workers, gloves, tools, cast-off hospital equipment, and excreta from rats and other laboratory animals used in radioactive tests). At a meeting with staff people from the Atomic Energy Commission on June 22, 1970, the company talked of "extremely favorable geological conditions in West Valley . . . such as seismological conditions, available disposable formation and impermeable cap rocks. . . ." Yet Government safety standards—and, with them, the costs of operation—kept rising; there were still no profits; and early in 1972, N.F.S. stopped all reprocessing (although the burial operation continued). In the announcement, the profit factor was hardly mentioned: Ostensibly, the operation was stopped so the plant could be decontaminated (radioactivity had gotten into places where it wasn't supposed to be), modified and expanded.

Expansion, however, required state and Federal permits, and that was opposed both by the State Attorney General, Louis J. Lefkowitz, and the then Conservation Commissioner, Ogden R. Reid. Lefkowitz filed papers with the Atomic Energy Commission in the fall of 1974 complaining of "an operation record which raises serious questions about risks to those who work there." Reid raised questions a year later about the radioactive wastes that had been dumped into Cattaraugus Creek, which is used for recreational boating and fishing. The permits were never granted, and the processing was never resumed. N.F.S. proved a failure, environmentally and economically, before it reached the ripe old age of six. Yet little of this was reflected in the company's public posture. An N.F.S. booklet issued in May 1974 still spoke in optimistic terms. "Reprocessing," the booklet instructed the public, "goes on inside concrete cells and 'canyons,' which keep the material locked securely inside. Air locks, absolute filters, extensive monitoring equipment and automatic safety systems simply prevent dangerous consequences—even in the case of equipment failure or human error." This despite all the cases (unknown to the public) in which the radioactivity did not remain "securely locked" and filters were not "absolute."

In the spring of 1975, the burial of radioactive trash was also discontinued, and on September 22, 1976 the company issued a press release announcing its decision "to withdraw from the nuclear fuel reprocessing business." Because of changing regulatory requirements, it said, the company would have to spend $600 million—almost 20 times the original cost—to make the plant viable. N.F.S. president Ralph W. Deuster said "the single most overpowering regulatory change was a drastic increase in the seismic criteria for the West Valley site which created doubt over whether or not the plant could ever be licensed for [expanded] commercial reprocessing operations."

SEISMOLOGICAL MISTAKE?

An admission of seismological error? How could geological conditions described by the company as "extremely favorable" in 1970 be cited as the principal adverse factor in 1976? Had there been some breakthrough in seismic knowledge during those six years? The seismological records were no secret. The fact is that between 1840 and 1967 there were 13 earthquakes, with epicenters within 100 miles of the N.F.S. plant, carrying intensities of 5 or higher on the Modified Mercalli scale. The Battelle Pacific Northwest Laboratory, under contract to the United States Government, conducted a study in 1976 and concluded that the West Valley site could have an earthquake with an intensity of nearly 8 once every 750 years or so. That may seem like a long time. But some of the nuclear wastes buried in West Valley will remain dangerous for hundreds of thousands of years. The Modified Mercalli scale is different from the more familiar Richter scale, but at a level of 7, furniture breaks, chimneys come down, plaster cracks, small slides occur, and large bells ring by themselves. In the case of West Valley, it means, at the very least, a transfer of risk to future generations.

THE CASE OF GERALD BROWN

An example of another kind of unforeseen problem may be found in the case of Gerald Brown. Brown, who is now 22, was one of the more than 1,100 young people who worked for N.F.S. as temporaries under the "body bank" concept. Basically, that concept is that your body is a bank and that each time it receives an intake of occupational radioactivity, some of the "bank" is used up. The young people had large "banks" to offer. Most had never worked in the nuclear industry before. The company needed them to do "hot jobs"—handling highly radioactive materials for short periods of time. It could not use its regular staff for this, since they had already received occupational radioactivity doses in their normal work. Upstate New York is a high-unemployment area, and the job offers delighted the young people. Some of them worked for only five minutes, were paid for the whole day, and were then let go.

Gerald Brown worked at these "hot jobs" from July 17 to Sept. 24, 1972. The radiation levels he received did not violate any Federal guidelines. But now he and his wife have had two sons suffering from Hurler's Syndrome—an incurable, terminal disease that is marked by dwarfism, retardation, failing eyesight and grotesque facial changes. The disease is rare and its causes are genetic: Both parents must have the recessive gene. Neither Brown nor his wife, Susan, know of any cases of Hurler's Syndrome occurring before in their families. That doesn't prove anything: Medical researchers say the syndrome may not recur for generations, and the Browns may not know if an ancestor had it in the 18th or 19th centuries. But the question nags: Was the disease caused by a genetic mutation induced by radioactive exposure? The company says there is no proof that Brown's work caused such a mutation, but Dr. Irwin Bross, director of biostatistics at the Roswell Park Memorial Institute in Buffalo, a cancer research facility, states: "The company can't say for sure whether this was caused by their genetic backgrounds or by radiation. Both are possible, and we can't be sure which. We can say, though, that the nuclear industry didn't understand what they were getting into and don't know how to get out of it."

Are there other illnesses that have been caused by the reprocessing operation at West

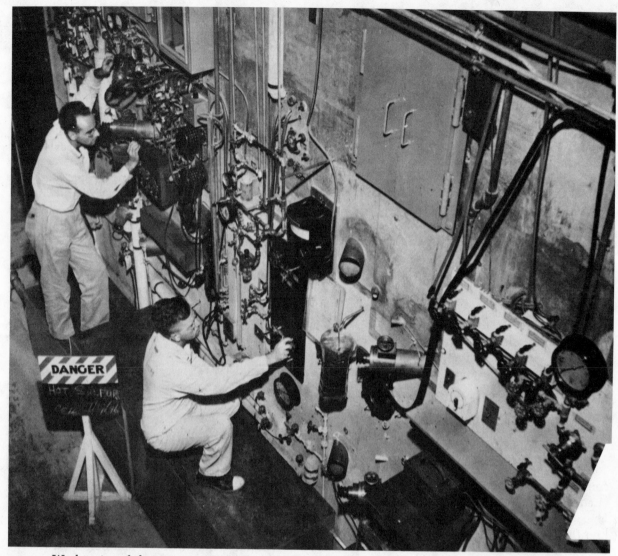

Workers in a laboratory in 1946 where the potentially dangerous material required the use of shields of lead or concrete and operation of equipment by remote control.

Valley? Nobody can say for sure; there are no hard medical data. The company is not required by state or Federal law to do follow-up studies on its employees and has declined, thus far, to give Irwin Bross and others facts about radiation levels received by former workers. Local medical authorities are reluctant to discuss so sensitive an issue. "I have practiced in this area for seven years," said Dr. Reza Ghaffari of Springville, "and my impression is that the number of congenital anomalies is high for the number of people who live around here. We have had lots of cancer, lots of hydrocephalics, lots of cleft palates. But this is just my impression based on what I know of the area. It is not possible to make any broader statements; the statistical information is simply not there."

INCIDENTS

But much information—of specific occurrences at West Valley—is there. The investigation has turned up at least 400 such "incidents." Here are some of them:

☐ The case of the improperly venting stack. It happened on June 11, 1968, and the radioactivity was spewed not only into the lunchroom but into the lobbies and the second and third floor windows and onto the front lawn. A report in the A.E.C.'s files says that both alpha and beta radioactivity were found on lunch tables and vending machines, that about 80 percent of the people in the lunch room had "hand and foot activities of between 10,000 and 50,000 counts per minute as measured with the hand and foot counter," and that some plant personnel inhaled radioactive particles. The lunch room was "decontaminated the next day." Apparently there was no way to decontaminate the front lawn, so it was literally dug up, taken to a radioactive burial site and buried.

Why the stack vented improperly is not explained satisfactorily in any reports found. The night supervisor's log said only that the "7D-13 tank had just finished jetting. The wind had swung from the east to out of the south." And the company's constant air monitoring charts, A.E.C. inspectors complained, showed no indicator of any "particulate emission" on the evening of June 11, 1968. Why? Government inspection records say that "Mr. Wenstrand [the plant's Health and Safety Manager] believes that there is a small 'pip' on the chart that coincides with the emission, but similar 'pips' have occurred before when there was no particulate emission, *so N.F.S. personnel would not have been alerted to a possible ventilation stack problem, on the basis of [monitoring] charts.*" (Italics added.)

□ In August 1966, a driver left his acid truck unattended. The truck rolled down a little hill and crashed into the south wall of the utility room, causing yet another week's shutdown and badly frightening a company supervisor who was using the nearby men's room when the crash occurred. He ran out of the men's room, tugging at his trousers, apparently convinced that the boilers had exploded and that West Valley was having its own little Doomsday. General manager William Oldham (not the executive in the men's room) recalls the incident. "It never should have happened," he said. "That damned fool forgot to chock his wheels."

□ The case of the worker with contaminated hair who was told to have a haircut outside. The incident occurred in March, and it remains unclear if the company considered what effect the recommended haircut might have on the barber, his instruments and his other customers.

□ On at least two occasions, railroad workers who coupled cars to locomotives on N.F.S. grounds found that their gloves had become contaminated with radioactivity. N.F.S. personnel confiscated the gloves and paid the men around $3 each so that they could buy new gloves. According to A.E.C. records, a brakeman asked plant personnel if the same contamination problem would persist in subsequent switching operations, "and N.F.S. personnel said, in effect, 'how should I know?'"

□ On April 14, 1968, a worker's finger was pricked by a plutonium-contaminated needle. The needle went through his fingertip. Inspection records say "the wound was surgically laid open and the tissue excised." The records do not say where the man is now or what the state of his health may be. Odds are he is fine, given the latency period associated with radioactive exposure. But no known effort is being made to follow his medical history and relate it to his employment.

□ In July 1969, it was decided that a device called a dissolver had outlived its usefulness, and company officals decided to entomb it "within the silty till in the high activity burial area." The dissolver was placed into a burial cask. En route to its final resting place, "a limited amount of contaminated solution dripped from the burial cask onto the roadbed." Moving with dispatch, management ordered that the dissolver, cask, and flatbed trailer, along with large chunks of contaminated roadway, would all share a common grave in the silty till.

□ Another Government report for July 1966 notes that workers found a way to chemically remove the yellow paint from the toes of the workshoes that had been provided by the company. The company had painted the shoes yellow because it did not want the shoes removed from the plant and its employees tracking radioactivity all over Cattaraugus and Erie counties. According to the report, company officials found out about the practice of removing the paint but did nothing to prevent it, "and now most personnel freely wear the supplied clothing [shoes] to and from the N.F.S. facility."

□ According to several former workers, some employees stole or "borrowed" highly radioactive tools from the plant and either sold them or used them in their own homes in the surrounding countryside. It remains unclear as to how they managed to get the tools past the plant's control system; where the tools are now; what, if any, radioactivity they now carry; and what effect this may have had on reports of an elevated rate in birth deformities in the area. No attempt is under way to find out. On at least one occasion, radioactive tools were sold at a local auction, and on another occasion a radioactive pump was sold at an auction in Pennsylvania. Where the stuff is now, nobody knows.

☐ Early in 1967, according to Government inspectors, "a truck driver, who was unloading acid at the storage area east of the plant, lost his shoes and pants because they became contaminated and could not be decontaminated to off-site limits."

☐ According to former employees, there were numerous leaks in the plant during the period of its operation. Many were not fixed. The dripping water was collected in plastic jugs. At least one jug had a sign on it that said "High Radiation." When it was filled up, the jug and its contents were supposed to be buried—like the roadbed, the dissolver, the flatbed truck and the front lawn—in the silty till. "We forgot about it once and the stuff ran all over the place," said one worker.

☐ On several occasions, the wash water in the laundryroom backed up, with the result that the floor of the room became contaminated. On Jan. 8, 1967, a company executive wrote that "initial counts for the day sample indicated high alpha activity" in the laundryroom. "Laundryman was taken out of area and area posted as high airborne pending long-lived alpha count." Two days later, the executive wrote that the "floor in laundry reads greater than 200,000 small d.p.m. [disintegrations per minute] alpha direct."

☐ On Feb. 9, 1967, employees Walt Zefers and Haafez Saadeck were ordered to evacuate the laundry area because of high airborne alpha radiation. In April of that year, Zefers quit N.F.S. Now, at the age of 68, he suffers from Paget's disease, which a medical dictionary defines as "a generalized skeletal disease of older persons of unknown cause, leading to thickening and softening of bones, as in the skull, and bending of weightbearing bones." On April 19, 1972, some five years after he inhaled "high

airborne alpha activity," Saadeck died of lung cancer.

Neither case can be attributed with any certainty to employment at N.F.S. In Saadeck's case, for example, his medical history shows that prior to working for N.F.S. he had been employed for at least 20 years in a forge and, before that, in a coal mine. He also smoked.

The company says it attempted to determine if radioactivity played a role in producing cancer in his lungs. However, the autopsy report on file in Buffalo's Millard Fillmore Hospital does not indicate that an alpha monitor was passed over Saadeck's lungs to determine if alpha activity was still present there. In any event, the lack of more thorough follow-up studies has made both cases the cause of speculation that may or may not be warranted.

☐ On one occasion, because an improperly trained worker did not turn on a valve, highly radioactive water went through regular drainage pipes that were designed to be used only for nonradioactive water.

☐ A truck driver made the mistake of walking through a puddle of uranium nitrate in a pair of expensive cowboy boots with fancy stitching. The boots, like the front lawn, the roadbed, the dissolver, the flatbed truck, the gloves, the pants and the jugs ended up in the silty till.

☐ The case of the unexpected visit to the home of a plant worker during Christmas week, 1969. The N.F.S. representatives who made the visit confiscated one pair of boots, one pair of workshoes, one fitted sheet, one baby blanket, two small throw rugs, one mattress cover pad, one footrest cover, one living-room rug, a pair of pants and a pair of socks. Some of these things were decontaminated and returned to their owner, while others went to the silty till. N.F.S. issued check 22011 for $26.70 for the boots and check 21939 for $97.43 for the living-room rug. It remains unclear how long the baby blanket and other items contaminated by something the worker brought home from the plant had been in the home before the confiscation.

☐ After a Christmas party, two workers showed up for the night shift drunk and slept through their shifts. Said one, "We were not in the best of shape to be around nuclear materials." At least one of the temporary workers admitted he got "stoned" on marijuana so that he could cope with his job. Another man who had a supervisory job had a reputation for coming in drunk with some frequency, even when it wasn't Christmas.

☐ In May 1966, according to A.E.C. reports, the low-level waste evaporator "experienced burping." A.E.C. inspectors said "the control system was found to be unsatisfactory."

☐ According to a former worker, the plant at one point hired a security guard who was an older man and had retired from work elsewhere. He worked nights in the gatehouse. "He used to leave the door open on hot summer evenings," the employee recalled, "and one night he saw a fox run by. He drew his revolver, fired, missed the fox but shot a tire that was on a parked car." The incident created a disquieting feeling among workers that an untrained security guard might send bullets in almost any direction in a complex containing so much radioactive waste.

LAPSES IN JUDGMENT

None of this means that the West Valley plant was designed or dominated by incompetents or that all of its shortcomings should be blamed on the engineers and physicists involved in the planning. It does suggest, however, that even the best planning and administration available could not cope with ordinary lapses in human judgment. The problem is what Dr. Bross calls the "failure in the interface between humans and hardware that engineers leave out of their calculations" when they produce statistical predictions and probabilities as to how a given plant is going to operate.

Dr. Marvin Resnikoff, a physicist and technical adviser to the Sierra Club, put it another way. "The position of the Sierra Club and mine personally is not unalterable opposition to nuclear power. . . . However, my conclusion is that the reprocessing industry is not yet mature, that numerous unresolved problems remain, not all of which are design problems, and that the requirements of the reprocessing industry are on the threshold of our technological and human ability."

Apparently, the threshold of governmental ability was also reached. William Oldham

Physicist Dr. Marvin Resnikoff

complains that N.F.S. tried for two years to get permission from the state to pump "some unanticipated water accumulations" that had been detected in the trenches containing radioactive trash. The company knew that something had gone wrong and wanted to pump the trenches dry before radioactive waste oozed out of the ground. The letters in his files indicate long delays and lack of decisive action on the part of the Bureau of Radiological Health in the State Department of Health. Finally, in March 1975, with the trenches overflowing, a plant official called the state in exasperation and said: "This is an emergency: We *have* to pump." The waste water was pumped into a treatment system, but—shades of Catch-22—they still didn't know what to do with it, and so it was discharged after treatment into Cattaraugus Creek. Oldham says the radioactive water was diluted by a factor of 2,000, and he feels that the toxicity of the stuff finally dumped into the creek was "negligible."

RISE IN MORTALITY RATE

What, then, is the legacy of the mistake at West Valley? Dr. Ernest Sternglass, professor of radiology at the University of Pittsburgh Medical School, says he calculates on the basis of Federal statistics that the infant mortality rate in Cattaraugus County rose 54 percent in the year after Nuclear Fuel Services began reprocessing nuclear wastes. Dr. Sternglass believes this was caused by small amounts of radiation contaminating water and milk produced within a 25-mile radius of the plant and shipped to the New York City area.

William Oldham, who feels that much of the criticism of his plant is unsubstantiated, was asked to recommend a scientist he felt could present an unbiased view of nuclear reprocessing generally. He suggested Dr. Bernard Cohen, a nuclear physicist who is also at the University of Pittsburgh. Cohen is widely regarded by both nuclear proponents and opponents as a man who sincerely believes in the future of nuclear energy—so much so that he has traveled far and wide at his own expense to explain his views.

Dr. Cohen said Dr. Sternglass's conclusions were statistically unsupportable. He affirmed his belief that plutonium is so safe that "I am willing to eat eight-tenths of a gram of it before a public audience." Dr. Cohen also said he would be willing to eat a smaller amount of strontium-90.

Nuclear technologists remove a reactor fuel element from a "pickling" vat. These fuel elements, made of radioactive metals, form the core of some reactors.

Anthony Roisman, an environmental lawyer in Washington who has worked closely with the Natural Resources Defense Council, says the idea of eating plutonium is "irrelevant," though he notes that there was a time when "people drank DDT to prove that it was all right." He has presented evidence to a review being conducted by the Federal Government to determine if there should be—and can be—a nuclear reprocessing industry in this country. Roisman notes that plutonium is relatively easy to handle but that once it becomes a powder, which happens during the reprocessing, it becomes far more hazardous, because it can be inhaled. It is lethal when ingested that way. Nuclear proponents, he says, are always asking for data *proving* that nuclear reprocessing causes harm, but "they know that the cancers will take 15 to 20 years to manifest themselves, and how will they be traced to nuclear reprocessing? The industry will probably blame cigarettes or color television sets."

Elsewhere, there is growing concern about nuclear fuel reprocessing, some of it caused by the West Valley experience, some of it because of technical problems that continue to escape solution. At Morris, Ill., where General Electric spent $65 million to construct a nuclear waste reprocessing plant, the company decided to mothball the facility before it processed any radioactive material. "We did not think it would be prudent to make the plant radioactive," said Dr. Bertram Wolfe of G.E.'s Nuclear Energy Programs Division. He added that because of changes in Government regulations, the prospect for ever using the plant is "low."

At Barnwell, S.C., a $250 million plant has only recently been constructed by Allied Chemial Company and General Atomic, a subsidiary of Gulf and Royal Dutch Shell. The Nuclear Regulatory Commission has not yet granted the plant a license to operate, and interveners before the commission have made it clear they don't think it should ever operate. The story is similar to what happened at West Valley: Rising Federal safety standards appear to require the plant to invest still more money, and this the company seems unwilling to do.

The whole arithmetic of the reprocessing industry is not what people thought it would be. In 1970, N.F.S. contracted with Consumers Power Company of Jackson, Mich., to provide reprocessed fuel for $22,400 a metric ton. Last July 13, however, N.F.S. told its customers that if it did any more reprocessing—and it wouldn't—the price per metric ton would have to be $1,009,300, an increase of 4,300 percent; a lawsuit is in progress. And the Government's experience in reprocessing this waste has apparently not produced the expertise the private sector would need to make reprocessing both safe and profitable.

WHO CLEANS THE SITE?

Urgent questions are also pending as to who will clean up the mess at West Valley. Under the original contracts, the bill for decontaminating and decommissioning the plant, and for the containment and ultimate disposal of the wastes, lies with the taxpayers of New York State. George W. Cunningham, director of the Division of Waste Management, Production and Reprocessing of the United States Energy Research and Development Administration (ERDA), said at a recent Congressional hearing that the "ballpark figure" for decommissioning the plant might be as much as $1 billion, although it could be "more on the order of $500 million to $600 million.... It depends on what one wants to do."

Actually, nobody knows what to do. A U.S. Nuclear Regulatory Commission report said in January that "no commercial or major ERDA [reprocessing] site has been decommissioned to date" and that "national standards for these aspects have not been developed." While the problem is being considered—and that may go on for years—New York State will have to pay between $2 million and $3 million annually just to make sure the plant is containing the radioactive wastes. New York is asking for a Federal bail-out. The state's Energy Research and Development Authority, in a report presented by its chairman, Dr. N. Richard Werthamer, said the problem is well beyond New York's financial and technical resources. It concluded: "A program which excludes West Valley from Federal ownership and control while including all other radioactive wastes would be an unjustified anomaly." As things now stand, the contract under which N.F.S.

The entrance to an underground laboratory where radioactive material has been used.

operated runs out Dec. 31, 1980, and the whole problem legally becomes New York State's alone.

THE INDUSTRY IN TURMOIL

"There is considerable doubt over whether there will be any further commercial reprocessing in the U.S. The reprocessing industry is in a state of turmoil..."

This admission of defeat came from Nuclear Fuel Services itself, in an affidavit filed with the United States District Court in Buffalo. What would a final abandonment of commercial reprocessing mean? It would not mean the end of the atomic power industry. It *would* mean that our uranium deposits would be used up that much faster, although there is now some difference of opinion as to how much uranium we have. Moreover, it should be borne in mind that reprocessing, even if successful, would not solve the nuclear-waste problem, since reprocessing generates wastes of its own. In any event, nuclear proponents are moving ahead with construction of atomic energy plants, feeling that if the spent fuel rods cannot be reprocessed, they can be stored. But where? Consolidated Edison has adequate space at its Indian Point plant to store spent fuel rods until about 1985, a company spokesman said. The rods are now in large, deep pools that resemble swimming pools. The water over the rods is supposed to act as a shield against excessive radiation leakage. If no commercial reprocessing or burial sites are available when Consolidated Edison runs out of space, "we'll seek more storage space," the spokesman said. The company is considering building more nuclear plants in the mid-Hudson area, about 90 miles north of New York City. But such plans are running into strong opposition from nuclear opponents and their future is uncertain.

Late in his Administration President Ford expressed his reservations about reprocessing, and President Carter has indicated his concern about what is perhaps the most powerful argument of all against commercial reprocessing: that if the United States develops such an industry in private hands, there would be a proliferation of such facilities all over the world, with the danger that any country owning reprocessing plants could obtain enough plutonium to make bombs. So, what to do with nuclear waste? The problem is all the more awesome in view of the atomic waste generated by the Government's weapons program. Government sources have estimated that the Federal establishment has accumulated between 75 million and 80 million gallons of highly radioactive waste since the end of World War II.

Present thinking among many scientists is that all stored liquid wastes, including those at West Valley, should be solidified—using one of several processes under development—and disposed of. Disposal suggestions under consideration thus far have included burial in a huge salt dome, far below the surface of the earth, with the hope (bolstered by extrapolations and probabilities, of course) that the burial area will not be disturbed by water, meteorites, bombs or earthquakes for hundreds of thousands of years ahead. Also proposed is burial of nuclear wastes in either Antarctica or Greenland, or disposal in the seabed of the deeper areas of the Pacific Ocean. There is another proposal: that all the world's radioactive garbage be loaded onto rockets and sent into high orbit around the earth, or around the sun, or into the sun. This is called "extraterrestrial disposal." Rockefeller, still firm in his faith in technology, hopes one of these methods will provide a solution for West Valley. He is sure that if rockets were used— and thousands of them would be needed to solve the problem on a national scale—there would be no misfires, that a rocket would not land in Cleveland or Yugoslavia by mistake.

As to what happened at West Valley, he says that if he were still Governor, he'd order an investigation. "But," he emphasizes, "you can't have a riskless society. Man's ability to contain, to channel and master his discoveries are what has made civilization." Asked how he would respond if another reporter came to him at some future date because all the reprocessing and disposal methods now under consideration had proved to be a mistake, Rockefeller replied: "I would say that we worked within the best knowledge of our time."

The Elephant Seal

by Walter Sullivan

The elephant seal, once near extinction after hunters slaughtered it indiscriminately in the last century, has now become such a prolific breeder that it is colonizing the beaches of California. The reproductive success of the three-ton animals, which are now estimated to number 65,000, forces upon conservationists a difficult decision: Should a protected species brought back from near annihilation be allowed to proliferate indefinitely?

Originally, northern elephant seals by the thousands bred and were hunted for their oil on beaches, from Point Reyes, north of San Francisco, to Cabo San Lázaro near the southern tip of Baja California. However, from 1884 to 1892, according to Dr. Burney J. LeBoeuf of the University of California at Santa Cruz, there is no record that a single elephant seal was seen anywhere.

Then, in 1892, specimen collectors for the Smithsonian Institution found eight of the huge animals on Isla de Guadalupe, which rises 4,800 feet from the Pacific Ocean west of Baja California. They killed all but one. While a few others were presumably at sea, the total number of survivors may not have exceeded 20 and the next generation may have been sired by only one or two dominant bulls.

That, according to Dr. LeBoeuf, was one of the most drastic "genetic bottlenecks" known. As a consequence, the thousands of northern elephant seals alive today appear almost as alike as identical twins. They therefore seem to lack the genetic diversity that might enable the species to cope with new diseases or other environmental changes. Dr. LeBoeuf fears that such genetic impoverishment may be faced by other species brought back from near extinction, such as the black-footed ferret, California condor, sea otter and California gray whale.

The northern elephant seal, weighing two to four tons and 16 feet long, is closely related to the slightly larger southern elephant seal, which inhabits the islands around Antarctica. It, too, was decimated by hunters of the last century, but apparently did not pass through so tight a genetic bottleneck and has retained more of its diversity.

The northern elephant seal began recovering initially because its sole rookery was too isolated

The elephantlike proboscus identifies the gender of this elephant seal as a male of the species.

to tempt hunters. When 264 of the animals were found on Guadalupe in 1922, Mexico outlawed a resumption of hunting, and the United States later followed suit.

The giant seals prefer islands but are confined to those with beaches, since they are too massive to climb onto rocks. In the 1970's, they have established themselves on two additional islands off California (Southeast Farallon Island off San Francisco and San Clemente off the town of that name) and two off Mexico's Baja California (Isla Natividad and Isla San Martin). Now, however, they have run out of islands within their habitat and in 1975, for the first time in this century, a female was observed to have given birth to a pup on the mainland—a beach 19 miles up the coast from here opposite Año Nuevo Island, whose beaches had become saturated with elephant seals.

Such is the rate of their population growth that last year about 100 pups were born on the main-land beach where five years ago there was but a single birth. Another 1,100 were born on the island. Some 570 adults occupied the mainland beach close to where Route 1, California's coastal highway, skirts the shore.

There are reports that elephant seals are already scouting the beaches at Point Reyes, the original northern limit of their range, but this could not be verified.

Dr. William Doyle, director of the university's Center for Coastal Marine Studies, foresees an increasing number of episodes in which Californians, awakened by deep bellowing, call the police to announce that "there is this monster outside my beach window—dying." Sooner or later, he fears, one of the huge animals will haul itself onto Route 1 in the dead of night "and some small foreign car will be demolished."

The massive animals hug the ground and often move so slowly they appear to be incapacitated.

THE ELEPHANT SEAL

They can, nevertheless, inflict a severe bite and the bulls, in their battles for dominance of a harem, sometimes inflict fatal wounds on one another. It is the males that carry the elephantlike proboscis that gives the species its name. The seals are docile unless approached within eight to 10 feet—a characteristic that helped expedite their near annihilation. However, there has been concern that for this reason tourists might come dangerously close.

□

For more than a decade Dr. LeBoeuf has tagged or marked 12,000 of the animals at rookeries from Guadalupe to the Farallons. They are painted with long poles, to provide a margin of safety, or with paint pellets fired by a gun. He and his students also have studied the extraordinary milk production of the females. After conception, the female spends close to a year at sea, fattening on squid, octopus and fish. In about mid-December, she hauls out onto a beach, joins the harem of an imposing bull and six days later drops her single pup. For the next 28 days, the mother nurses her pup, which gains weight at a phenomenal rate. In those four weeks, it increases from 60 to 300 pounds. And for every pound it gains, the mother loses two, since she neither feeds nor drinks in this period.

By means of a suction device at the end of a long pole and other methods, milk samples have been collected and analyzed. The mean level of their fat content is 54.5 per cent, the highest found in any mammal, and the water content is only 32.8 per cent, the lowest known. The meager water content presumably helps the animal produce so large a volume of milk without drinking.

In the last four days of the nursing period, the mother mates with the bull, or an interloper, and then returns to the sea, leaving the pup to molt and learn to swim on its own. Life for a pup in the rookery is hazardous. They are often trampled and crushed in the efforts of the bulls to achieve high social rank and copulate. According to Dr. LeBoeuf, "Males are impervious to a pup's presence and neither its shrill cries nor its mother's aggressiveness persuades a two- to three-ton bull to move when he comes to rest on a pup."

Dr. LeBoeuf has found that 13 to 14 percent of the pups on Año Nuevo die before they are weaned. Some, when they become separated from their mothers, try to suckle other females. Often, in response, they are bitten or have their skulls crushed. Incited by the pups' squawking, five or six females may "mob" it with fatal results. However, a third of the mothers that lose their pups adopt another.

Before settlement of the West Coast, the elephant seal population on the mainland was kept down by such predators as grizzly bears and cougars. But such checks no longer exist. The giant seals have become a tourist attraction at Año Nuevo Point, where an estimated 245,000 visitors have come to see them in the first three years after their arrival. Sooner or later, however, a confrontation seems inevitable between elephant seals and such users of California's beaches as surfers, nudists and conventional bathers.

BIOLOGY

If the "hunting of the quark" is physics at its most basic level, then dissecting the "code of life"—the messages written into DNA—is its biological counterpart. DNA acts, in controlling life processes, much like the punched tape used to control certain computers and other devices. Perforations in the tape form patterns analogous to letters in a word and a certain number of such letters carry a specific piece of information.

In DNA the "letters" are clusters of atoms (nucleotides) that link to form the famous twisted structure, or double helix, of the DNA molecule. The alphabet of DNA is extraordinarily simple, considering that all of our inborn characteristics, from hair color and sex to the most minor chemical function of the body, are coded into such a molecule. There are only four "letters" to the alphabet, usually represented as A, G, C, and T (for the nucleotides adenine guanine cytosine and thymine).

In February, 1977, researchers at Cambridge University in England created a sensation in the world of biology by publishing the full sequence of 5,375 nucleotides forming the DNA of the bacterial virus Phi X-174. By then researchers at

Yale University had almost completed the sequencing of SV-40, a virus that causes cancer in monkeys. An 800-link sequence in its message seems peculiar to that virus, hinting at the possibility that it may be related to its cancer-causing properties. Several analytical methods are being used, but they all have roots in those developed over the past thirty years by Dr. Frederick Sanger at Cambridge. He first applied them to proteins—those building blocks of all living matter—which are long chains of amino acids. The sequence of those acids is determined by the sequence of "letters" in DNA. It was for his work on proteins that Sanger won a Nobel Prize in 1958. Now, taking advantage of the key-and-lock manner in which DNA does its job, he has applied similar methods to that molecule. In nature the reading out of the DNA message—transferring it to other molecules and eventually into the manufacture of proteins—depends on a special property of the four letters of the DNA alphabet. Each is receptive to one, and only one, matching substance, like a lock that accepts only one kind of key. The DNA molecule, in the presence of such matching substances, can therefore assemble a sort of chemical mirror image of itself—a messenger to carry its information into the protein-forming machinery of a living cell. By breaking up the DNA and then exposing the strands to matching substances that have been tagged radioactively, and by a variety of other chemical tricks, it has been possible to sequence the entire structure of viral DNA. An amazing discovery has been the fact that rather long parts of the message overlap. It is as though by reading a passage from a book beginning at the top of a page, you received one meaning, but by starting to read a few letters past that point, an entirely different message came through.

Obviously a tape recording with sufficient information to program the function, structure and reproduction of a higher organism such as a giant tree or human being—must be extremely long, with millions or even billions of nucleotides in its chain. Yet ways have now been developed to determine the sequence of bases in the chain—to "read the code of life." For a couple of the simplest organisms, such as a virus that infects bacteria native to the intestine, the full length of the molecule, involving thousands of nucleotides, has been deciphered, or "sequenced." For higher organisms it seems more practical to sequence key sections of the molecule—those bearing on some critical part of the life process or those that have been manipulated by "genetic engineer-ing"—rather than attack the entire molecule, with its billions of nucleotides.

The research in this field is also uncovering signals that start and stop key portions of the message, like paragraph indentations. One of the most tantalizing discoveries has been that it is possible to activate dormant DNA in a cell that had been performing a highly specialized function, controlled by only a tiny fraction of the DNA, the rest having been turned off. The work has seemingly confirmed that every cell nucleus of a person's body—millions of them—contains the DNA message that controlled his or her development from egg cell to adult. In the work of John Gurdon, which he is pursuing in the same laboratories at Cambridge where DNA is being sequenced, he has removed the nuclei from the specialized body cells of frogs, inserted them into eggs of other frogs whose nuclei have been destroyed, and developed frogs identical, not to the parent of the egg, but to the frog robbed of a single skin cell or intestinal cell. The implication is that many, if not all, cells in our bodies, including those sloughed off everyday, are potential individuals, complicating even further the ethical and biological problem of defining when "human" life begins in the gestation process. The focus of interest for researchers in this field, however, is directed at what "turns on" the dormant DNA. Why can a salamander grow a new leg, but a human being cannot?

The Wolf Gets a Better Image

by Boyce Rensberger

A crimson pool of blood lay frozen in the snow. Ravens pecked at the last shreds of what had been a deer. "There they are!" David Mech, wolf biologist, shouted over the roar of a Cessna 180 circling tightly above the frozen lake as he pointed down at the shore. "The wolves!" About a hundred yards from the kill in the middle of the lake, six timber wolves rested drowsily after the kill in the cold, bright sun. One jumped up to watch the little plane as it banked sharply.

Dr. L. David Mech, who has studied wolves for 18 years and is widely acknowledged as this country's leading expert on wolf behavior, marked down on a tracking form the location of the wolves, how many there were and what they were doing. Then he directed the pilot to head southwest toward the place where another wolf pack had last been seen. Through such aerial observations, a technique he helped develop, Dr. Mech has gathered much of the evidence that has debunked many of man's oldest myths about the wolf. Once widely hated and persecuted as a dangerous predator, the wolf today, thanks largely to Dr. Mech's research, is coming to be regarded as an ecologically important member of its wilderness habitats and as an animal with a complex and fascinating society. Once feared as dangerous to people, the wolf is now known not only as friendly and sociable within its pack, but as no threat to man. There is no documented instance of a free-living wolf attacking a person in North America.

Dr. Mech (pronounced Meech), now 40 years old, began his wolf research in 1959 as a graduate student while observing the two dozen wolves of Isle Royale National Park in Lake Superior. Today, employed by the Endangered Wildlife Research Program of the United States Fish and Wildlife Service, he is in charge of a wide-ranging, long-term study of the relatively stable population of 1,000 to 1,200 wolves in northern Minnesota. These animals are the last substantial population of wolves in the United States outside of Alaska. The greatest concentration of these wolves is in the Superior National Forest near Ely, Minnesota in the extreme northeastern corner of the state, and the dead of winter, when the beasts can be spotted against the snow, is the best time for studying them.

RADIO-TAGGED WOLVES ARE TRACKED

As the little plane headed toward the next wolf pack sighting, Dr. Mech put his headphones back on and listened intently for a

201

clicking signal picked up by antennas mounted on the plane's wing struts. One of the wolves in the pack, like one in the pack sighted near its kill, was wearing a tiny, battery-powered transmitter on a collar around its neck. The signal from this constantly operating device picked up by the antennas on the plane guides Dr. Mech to the wolves. Over the years Dr. Mech and his assistants, most of them graduate students in wildlife biology at the University of Minnesota, have trapped and radio-tagged about 140 wolves in the Superior National Forest area.

After being captured in a modified leg-hold trap, the wolves are immobilized with drugs, weighed, and identified by sex. Blood samples are taken. Ear tags are clipped on and the collar

The wolf is examined and measured.

L. David Mech attaches a radio collar for future studies.

is fitted. As the drugs wear off, the wolf, transmitting on its own frequency, runs off to rejoin its pack. Because wolf packs are stable social units, the signal from a single radio collar can lead Dr. Mech or his students to the entire pack. Because the antennas are highly directional, picking up signals from either the left or right side of the plane, depending on which antenna is used, Dr. Mech can tell where the transmitting wolf is in relationship to the plane. Dr. Mech can switch from one antenna to the other. When the signal is equally strong from the two antennas, the plane is headed directly for the wolves.

Twenty-four of the 140 radio collars are still working, the others either have stopped operating (usually after a year or so) or the wolves have been killed (wolves that venture too near human beings risk being shot or trapped). The 24 tagged animals represent nine packs, one newly formed pair that may breed to establish a new pack, and four lone wolves, animals who have left their original packs to wander alone and sometimes find a mate and a vacant "territory" in which they can establish a new pack.

Minutes after leaving the pack with the deer kill, Dr. Mech signaled the pilot to circle above a forested ridge. "They're down there," Dr. Mech shouted. "Can't see 'em. They're probably under the trees." After logging their position and that of some other packs, Dr. Mech headed back to the airport. Every day during winter and at least weekly during the summer Dr. Mech or his students go up in planes to find the collared wolves. One pack has been tracked for six years. When the locations for a given wolf pack are plotted on a map, almost all fall within a tightly circumscribed territory abutting the territories of other wolf packs and almost never overlapping them. One pack, for example, has fluctuated from two to nine members over the years, but has always maintained the same territorial boundaries with its neighboring packs. On a larger scale, wolf densities usually are about one for every 10 square miles. Wolves mark the boundaries with urine and, even when chasing prey, seldom enter alien territory. When they do, they risk attack from the resident pack.

WOLF BLOOD AND BEER

Back at the log cabin on a Forest Service compound near Ely that serves as the wolf study's field station, Dr. Mech and half a dozen students come and go throughout the day, drying out soaked gloves, pouring hot coffee, calibrating radio receivers, exchanging information on the day's sightings. In the refrigerator, vials of wolf blood share space with bottles of beer. Several of the students track wolves from the air and others manage projects on deer, lynxes, moose, snowshoe hares, ravens and other local fauna. Many of these animals are also wearing radio collars. By studying both wolves and their prey, Dr. Mech and his students hope to discover and understand those elements of their behavior that have evolved as ways of coping with the other species. Anatomical adaptations for attack and defense are well known but behavioral adaptations are not.

Biologists have long known that wolves have developed certain ways of hunting that maximize their chances of killing deer. Presumably deer, who heretofore have not been intensively studied as one-half of a predator-prey relationship, have evolved defensive behaviors as well. One deer behavior that is under study is the congregating of deer in open meadows, or deer yards, in the winter. In spring and summer deer are dispersed through the forest. Why do deer shift back and forth between two different systems of social organization? Somehow, Dr. Mech suspects, the deer's slower metabolism in winter, the difficulty of moving in snow and the fact that fawns have grown more independent since their birth make "yarding" a better way to defend against wolves in winter but a poorer way in summer. A four-year study of deer behavior is planned in an attempt to explain this phenomenon.

One recent morning a student reported that wolf No. 2407 and its pack were found well out of their territory. Dr. Mech checked the location on a map. "That's interesting," he remarked. "Those sons of guns, they're trespassing, really striking out on their own." From the map it appeared that they would have had to cross two roads to reach their present location from where they had been the day before. Later in the day,

*The migratory pattern of a lone wolf
as traced by Dr. Mech.*

after a futile attempt to find a radio-collared wolf that someone reported seeing on a road with a trap on its foot, Dr. Mech drove the snow-covered roads that the "trespassing" wolves had crossed to look for scent marks. When wolves cross a road or other physical boundary, they mark the junction with urine. He wanted to collect urine samples frozen into the snow for biochemical analysis but a snow-plow had recently obliterated the marks.

AN ANCIENT ANTIPATHY

The biologists' interactions with the human population in and around Ely have proven both rewarding and frustrating. Dr. Mech said that although most townspeople were sympathetic to the wolf research and favored the species' protection, a few retained the older antipathy. There is a vigilante group that kills wolves whenever possible and puts the carcasses on other people's doorsteps with notes arguing that wolves destroy deer that should be protected for hunters. Although wolf hunting and trapping have been illegal in Minnesota since 1965, it continues. Whenever one of the collared wolves is killed, however, some sympathetic trappers notify Dr. Mech by leaving anonymous notes at a local bar. Because some of Dr. Mech's colleagues study deer, some townspeople are puzzled that scientists who specialize in species that are seen as "natural enemies" can be friends. "We have a ways to go in changing peoples attitudes about these animals," Dr. Mech said.

Dr. Mech says he likes to get into the field as often as possible but noted that he had a desk job with the Fish and Wildlife Service in St. Paul. There, with access to libraries and laboratories, he writes scientific papers on wolf behavior, consults with scientists and conservationists around the world. Last year he spent a month in India training biologists there in radio tracking. He also develops strategies for protecting wolves. For example, Dr. Mech was heavily involved in the 1974 effort to relocate four Minnesota wolves in Michigan's Upper Peninsula. Within eight months after the wolves were released, all four, two males and two females had been killed by human beings. Two were shot; one was trapped and then shot; one was hit by a car. The experiment did establish that relocated wolves could establish themselves in a new territory and survive. "The problem," Dr. Mech said, "is the human population. Next time we would want to do a more intensive public education effort."

Dr. Mech is a member of the Eastern Timber Wolf Recovery Team appointed by the Fish and Wildlife Service to devise a program for the protection and re-establishment of wolf populations. The group has suggested that wolves be reintroduced to wilderness areas in Michigan, Wisconsin, New York State's Adirondacks, Maine, and the Great Smoky Mountain National Park. Dr. Mech has found that wolves can double their numbers every year. However, they do not if the area is already full and each pack's territory abuts others on all sides. Lone wolves, unable to establish a territory near their place of origin, disperse to a less desirable habitat and, in many cases, are killed by people. Thus, Dr. Mech has found, wolf hunting or trapping can continue at a substantial rate on the fringes of prime wolf country without lowering the average wolf population. In Minnesota, for example, the Wolf Recovery Team has recommended that controlled wolf killing be permitted in a buffer zone around the wolf's 10,000-square-mile prime range, which would remain totally protected. Limited wolf hunting or trapping, the group believes, is necessary to minimize the loss of livestock to wolves and to increase the base of local citizen support for conservation, without which wolves might not survive at all.

Building a Better Bug Trap

by Anthony Wolff

There are several million distinct insect species, maybe 250 million individual bugs for every one of us. As far as human beings are concerned, about 99.9 percent of these are benign, and a few, like bees, are indispensable. But the rest, perhaps a couple of thousand species, are pests: Either they eat the same things we do, or they eat us.

So the practical question is not whether we are going to kill insects, but how. The entomologists—biologists who specialize in insects—insist that they know the enemy best, that they are most qualified to keep the insects under control without unnecessary ecological violence. But ever since the end of World War II, when DDT was mustered out for civilian use, to be followed by a whole generation of powerful synthetic poisons, the entomologists have been practically eclipsed by the chemists. For about 25 years, it was almost universally agreed that chemistry was the final solution to the insect problem.

The trouble with the poisons, a growing chorus of biologists has been warning since DDT's debut, is not only that they are dangerous to birds and other living things, including people, but that, in the long run, they don't work. Because of the sheer number of insects, the variability among individuals of any species and their prodigious talent for reproduction, sooner or later a roll of the genetic dice will bless them with immunity to any poison. In fact, several hundred pest species and strains have already developed partial or complete tolerance to DDT. More ominously, despite the chemists' best efforts to outrun insect reproduction with increasingly potent poisons, a number of important pests—including spider mites, which attack fruit crops, and some malaria-carrying mosquitoes—have become resistant to all known insecticides.

In 1972, 10 years after Rachel Carson's "Silent Spring" dramatized pesticide misuse into a public issue, the Federal Environmental Protection Agency finally got around to proscribing virtually all domestic use of DDT. Since then, despite unending protests from those who claim pesticide disarmament is leaving us wide open to insect attack, the E.P.A. has extended its ban to a number of other major poisons and imposed strict constraints on both old and new insecticides. These bans and constraints, and the increasing ineffectiveness and ecological ill-repute of chemical poisons, have diverted attention and support to the entomologists for the first time in a generation. But, ironically, though their most sophisticated new methods promise dramatic results, strict pesticide regulations established in DDT's aftermath have made it difficult for the entomologists to get their methods accepted for use outside the laboratory.

BIOLOGICAL CONTROLS

Instead of crude poisons, the entomologists have been insisting all along, there are better ways to commit insecticide. The alternatives they propose are generally characterized as "biological controls," to indicate that they rely on sophisticated tinkering with the pests' own biology and ecology. As a science, biological control emerged from folk entomology a century ago with a strategy of controlling pest insects by encouraging their natural enemies— predators, parasites and pathogens, known in the business as the Three P's. Today's suburban gardeners who mail-order praying mantises or ladybugs to devour their backyard pests are putting their faith in classical biological control.

Gradually, the definition of biological control has broadened to include a variety of insecticidal plots calculated to defeat pests without disrupting the rest of the ecological system. Biological control can be as simple as altering planting schedules to put a crop's life cycle out of phase with a pest's. Or it can be as sophisticated as the "sterile male" technique currently used in Texas and Mexico against a cattle pest called the screwworm fly. The active ingredient in this technique is millions of male insects that are artificially bred and sterilized by exposure to radiation. Released periodically, the sterile males overwhelm their fertile, wild cousins in the competition for females. The preponderance of fruitless matings depresses the population until, after several generations, the pest is practically exterminated.

The most sophisticated and potentially important of the new biological controls, currently nearing the operational stage after two decades of patient development by a small avant-garde of entomologists, have been christened "third-generation insecticides." Like first-generation materials—prewar standbys such as arsenates of

lead, nicotine and kerosene—and those of the second generation inaugurated by DDT, third-generation insecticides are potent chemicals, but there the resemblance ends. The new bug killers are not foreign to the insects, but man-made, near-perfect counterfeits of the insects' own chemicals, formulated by eons of evolution to carry out the pests' most basic life processes. Instead of poisoning their targets—and innocent bystanders—the third-generation materials infiltrate and sabotage the pests' own chemical communication systems apparently without affecting anything else.

JUVENILE HORMONES

The first of the new insecticides to be developed is a group of compounds called juvenile hormones, JH for short. Harvard's Carroll Williams, the leading figure in JH research, calls them "status quo hormones." Normally, during its wormlike larval stages— between its emergence from the egg and its metamorphosis into an adult—the insect's own juvenile hormone restrains its maturation. Precisely on cue, when the larva attains a specific weight, its supply of JH is abruptly turned off. Only then, free to obey the urging of a complementary growth hormone, can the insect begin the dramatic transformation into its adult form.

The source of the potent JH hormone, two tiny glands near the insect's brain, was located by the British biologist V. B. Wigglesworth in 1936. It took 20 years, however, for Harvard's Carroll Williams, a former graduate student of Wigglesworth's, to isolate the active substance in the abdomens of male *Cecropia*, the North American silkworm moth. Soaking the crushed abdomens of *Cecropia* in an ether solvent, Williams extracted an impure "golden oil" which he tested on silkworm pupae—dormant insects in transition between the larval and mature stages. "The hormone penetrated the unbroken skin," Williams reported, "and all individuals ultimately died without completing metamorphosis." "It seems likely," he predicted in 1957, "that the hormone, when identified and synthesized, will prove to be an effective insecticide." It took a team of researchers led by

Herbert Roller at the University of Wisconsin another 10 years to refine and identify Williams's golden oil. The pure hormone proved to be so potent that a single gram could theoretically kill a billion insects. In the effort to synthesize the complex JH molecule, researchers have since scored thousands of near misses, producing JH analogues less potent than the real thing but so close chemically that insects can't tell the difference. This imprecision seems to be effective.

The first practical juvenile-hormone formulation, trade-named Altosid, was registered for general release in 1974 by Zoecon, a California spin-off of Syntex, the company that synthesizes the human hormones used in The Pill. According to the manufacturer, as little as two ounces of Altosid per acre is sufficient to achieve complete control of certain mosquitoes that have become totally indifferent to DDT and all other poisons. Altosid also works on flies: Added to salt licks for cattle, JH survives the bovine four-part stomach and when excreted, controls houseflies and hornflies, major cattle pests that breed in manure piles.

Initially, juvenile hormones were greeted as ideal alternatives to poison insecticides. According to all the evidence, JH is fatal to many insects and innocuous to other living things. In addition, Carroll Williams hypothesized in 1956, "insects can scarcely evolve a resistance to their own hormone." While this theoretical assumption has since been challenged—on the ground that the JH analogues are not exactly the insects' own formulas, and that insects already have mechanisms for disarming their own JH at the right moment—no insect resistance to man-made juvenile hormones has as yet been evidenced.

Nevertheless, JH is not the perfect insecticide. For one thing, it tends to affect insects indiscriminately, the benign and the beneficial as well as the target pests. Also, most man-made JH compounds are notably short-lived in the environment, while the insect's susceptibility to them is similarly brief; thus, insect and insecticide must be brought together on a more precise schedule than can easily be arranged. Most important, JH insecticides are ineffective during the insect's larval stage, while the insect is already under the influence of its own juvenile hormone.

"This insensitivity of larvae is inconsequential in the case of most medically important insects, where the vector is usually the adult," Carroll Williams explains. "For example, a larval mosquito poses no problem as long as one can block its metamorphosis. But the situation is quite the reverse in the case of most agricultural pests, where the growing, feeding larva is usually the most damaging stage in the life history."

Early in his work on the juvenile hormone, Williams inferred the existence of another hormone, an "anti-JH," that must signal the insect's glands to stop producing JH at the critical moment. If such a hormone were found, Williams reasoned, it would be effective against larval insects by "provoking a precocious, lethal metamorphosis." Recently, after exhausting several generations of the graduate students and post-doctoral researchers who gravitate to his basement domain at the Harvard Biological Laboratories, and tens of thousands of laboratory-bred tobacco hornworms, Williams succeeded in isolating an active anti-JH substance. Then, earlier in 1976, another researcher, William S. Bowers, at the New York State Agricultural Station in Geneva, N.Y., announced the discovery of two natural anti-JH compounds in a common garden plant, the blue-flowered ageratum. Bowers's best guess is that the compounds are the plant's vestigial defenses against insect pests in its evolutionary past. Dubbed "precocines"—from "precocious"—Bowers's finds turn out to be simple molecules that can be mass-produced in a two-stage chemical process. Moreover, Bowers predicts that by fine-tuning the formulas, these anti-JH analogues can be aimed at specific target pests without affecting innocuous or beneficial insects.

While the JH and anti-JH compounds frustrate normal insect growth processes, another class of biochemical insecticides, called pheromones, sabotages the pests' programmed sex lives. Akin to hormones, which transmit internal messages, pheromones are potent perfumes that broadcast information among individuals of a species. Many insects—as well as higher animals—use specific pheromones to communicate about property rights, defense strategy, food location, the way home and other vital matters. More to the point, in many insect species pheromone communication is essential for sex, taking the place of flirtation and foreplay. Until he smells the female's sex pheromone, the male cannot find her. Worse yet, if a lady bug fails to emit the appropriate perfume, a feckless male can stumble on her and not have the slightest inkling of what to do.

Insect pheromone communication is one of nature's most curious and awe-inspiring technologies. A ready and willing female atomizes from the tip of her abdomen a millionth of a gram or less of sex pheromone, diluted to 200 molecules per cubic centimeter of air. A distant, downwind male, preoccupied by some other business, with sex the farthest thing from what passes for his mind, turns into the wind, wings his way toward the source of the scent and arrives ready to copulate. The champion long-distance lover of record is the male *Bombyx mori*, a Chinese silkworm moth that responded to a come-hither pheromone from a female who was 6.8 miles away. According to calculations by an entomologist given to pheromone fantasies, the same talent on a human scale could stampede all the men on the Eastern Seaboard to a single woman in Omaha.

INGENUITY TRAPS BUGS

Paradoxically, the very perfection of the sex-pheromone system makes it a potent insecticidal weapon. When an insect is under the influence of its sex pheromones, love is blind: The creature will make a pass at the source of the odor, whether it is the abdomen of a female moth or a lab assistant's elbow. The love-struck bug's inability to tell the difference has inspired a number of ingenious plots. In the simplest scheme, pheromone-baited traps are used to monitor pest population levels. The early-warning system allows a farmer to resort to costly, dangerous insecticides only when and where pest infestation is imminent, instead of following the usual practice of prophylactically spraying his whole crop on a fixed schedule. In experimental orchards at Cornell, Dr. Wendell Roelofs has used the system to save up to 50 percent of the usual cost of spraying against the red-banded leaf roller. Sex-pheromone traps recently detected the arrival in San Jose of the

Entomologist David Wood with bark beetles.

gypsy moth, a notorious leaf-eater from the East. California entomologists credit the technique with giving them a two-year head start on a major pest outbreak.

Dr. David L. Wood at the University of California at Berkeley is using sex-pheromone strategy to "trap-out" entire populations of the western pine beetle, one of the nation's most destructive timber pests. In 1975, the western pine beetle and its relatives were blamed for the loss of some 15 billion board feet of commercial timber, worth more than $900 million as lumber and enough to build 8,000 houses. Like many entomologists, Wood, 45, has devoted the better part of his career to a single insect family. Beginning when he was a graduate student, in 20 years of spying on bark beetles in laboratory and forest, Wood discovered that the mass beetle attacks that kill trees are invariably preceded by chance visits from single males. The male burrows through the bark into the tree's tender tissue, hollowing out a nuptial chamber. Then, as though on cue, three female beetles enter the chamber ready for mating, and the fatal population explosion begins.

By a process of elimination, Wood determined that the timely arrival of the females is prompted by a male sex pheromone in the "frass," the mixture of sawdust and excrement produced during the homebuilding effort. The delicate task of separating the pheromone from the frass, identifying it and then synthesizing it was turned over to a team headed by Dr. Robert M. Silverstein at the State University of New York at Syracuse. Silverstein has identified more insect pheromones than any other researcher; nevertheless, the western pine beetle took him almost three years.

BLENDED INSECTS

The basic procedure for pheromone analysis begins with as many as half a million insects that must be hatched, raised to maturity and slaughtered. Then their abdomens are reduced to a pulp in a blender, refined in a series of solvents, strained, filtered, washed and rewashed, all under strictly controlled conditions. From all this effort, the researcher gets a few drops of crude pheromone extract, still "dirty" with impurities. Finally, the extract is distilled

and assayed by several kinds of chromatographs, spectrographs and spectrometers. At each stage of the process, the material is bioassayed—tested on living insects—just to make sure that the expensive machines haven't made a mistake. If they have—and they do—the whole process, months of work, starts again from Step 1. Recently, researchers have learned how to "catch" some pheromones in midair with odor-absorbing materials, eliminating much of the mess of the bug-slaughtering classical technique and saving time. Still, pheromone chemistry is not a business for people with short attention spans.

Once a pheromone has been identified and synthesized from raw chemicals—another often-difficult process—it is field-tested in the pest's natural environment. For David Wood's western pine beetle pheromone, the test site was at Bass Lake in the Sierra National Forest of northern California. In collaboration with U.S. Forest Service entomologist William Bedard, Wood booby-trapped two separate square-mile areas with large pheromone-baited panels coated with a goo called Sticken Special. In a two-month period, the traps captured more than 400,000 pine beetles. The next year, mortality from beetle attacks dropped from 227 trees to 84. Encouraged by their Bass Lake results and other test data, Wood and Bedard are applying to the Environmental Protection Agency to register their pheromone trap-out technique for operational use against the western pine beetle. If their application is approved, as they expect it to be, theirs may be the first pheromone formulation to graduate from experimental status to practical use as an insect control.

Their closest competition is Dr. Harry H. Shorey at the University of California's Riverside campus. His preoccupation with pheromones began almost as soon as he graduated from Cornell 17 years ago, when only two pheromones had been identified and there was no such thing as a pheromone specialist. One day, investigating a flurry of activity in a cage full of moths, Shorey found that the nether end of a single female was the focus of a large number of males engaging in copulatory behavior. From the abdomen of another female, Shorey squeezed a tiny drop of pheromone onto a piece of paper, which immediately became as

Dr. Harry H. Shorey in his laboratory. He discovered that too much sex pheromone can be an effective insect repellent.

popular with the male moths as the first female had been. "Since then," says Shorey, "I haven't had another interest." From time to time, Shorey likes to speculate—and even experiment in a casual way—on the possibility of sex-pheromone communication among people. Most often, though, he thinks about the sex life of the pink bollworm, a major cotton pest.

TETHERED LURES

Shorey's strategy against the pink bollworm was inspired by a pheromone effect he discovered in 1967. Experimenting with the cabbage looper, another pest, Shorey placed virgin males and females together in closed jars in which the air was saturated with synthetic looper sex pheromone. Instead of the expected orgy, Shorey reported, "no males mated with

the females even though the confined conditions restricted all moths to within 15 centimeters of one another and numerous females were observed in the supposed receptive position, with sex-pheromone glands extruded and wings vibrating."

Shorey guessed that the males' pheromone receptors, numb from overstimulation by the ambient perfume, were unable to respond to the tiny increment of scent released by the females. Evidently, as a sex inhibitor, too much pheromone was as good as none at all. Could the same effect be produced in an open field instead of a closed jar? In a small-scale test, Shorey glued threads to the bodies of virgin female loopers and tethered them overnight in a field, pushovers for any male with a whim. Their only protection was a few milligrams of synthetic pheromone perfuming the night air from some nearby evaporators. When the sun came up the next morning, the virgins were still inviolate.

Switching to the pink bollworm, which had only recently invaded California, alarming the state's cotton industry, Shorey mounted a larger campaign under actual field conditions. Each week for 10 weeks, his graduate students tied 30,000 loops of string, impregnated each with precisely 10 milligrams of pheromone and distributed them over 12 acres of cotton in the Coachella Valley. During the growing season, traps baited with female moths in adjacent unprotected fields captured more than 11,000 males per trap while traps in the pheromone-treated fields averaged only 38 males each. At harvest time, each cotton boll in the untreated fields was infested with an average of 6.9 larvae, while the bolls in the protected fields were practically wormfree.

The only trouble with the experiment was that it depended on an unlimited supply of unpaid field hands. And, fortuitously, Shorey's chemical in search of a delivery system found a delivery system in search of a chemical. The hardware, which has a unit cost close to nil, no moving parts and no need for service, is simply a capillary tube—an inch or two of extremely fine, hollow, plastic fiber, like a miniature strand of spaghetti. Loaded with a pheromone, the fiber becomes, in accordance with the laws of physics, what its manufacturer calls a "controlled-release vapor dispenser."

Last summer, under an experimental permit from the Environmental Protection Agency, plastic fibers loaded with "gossyplure," the pink bollworm pheromone, were distributed over 2,700 acres of Arizona cotton. Even before the computer had time to digest the results of the harvest, Conrel, the Massachusetts company that makes the plastic fibers, was sufficiently encouraged to plan another commercial demonstration on 10 times as much cotton next year. According to Conrel's project manager, the need for poisons to suppress the pink bollworm was reduced by as much as 70 to 80 percent, and "the farmers want to do it again." "We have proved the point," he says. "Next year, we'll be fine-tuning the system."

A BETTER MOUSETRAP

So the pioneers in third-generation insecticides are beginning to see some recognition for their 20 years of research. Williams, Wood and Shorey shuttle to New York, Rome, Stockholm, Tokyo and points in between to satisfy the growing interest of fellow-scientists and policymakers. Financial support for their work is assured: Early seed money for each from the Rockefeller Foundation has been supplemented by important grants from such donors as the National Institutes of Health, other Federal and state agencies and industry groups. Still, after these same 20 years, only a handful of the new insecticides have made it even to the threshold of practical use. By nature and by nurture, entomologists tend to be compatible with the purposeful pace of research, the measured cycles of academic years and insect generations, the slow progress of science. Now, however, they believe they have built the insecticidal equivalent of a better mousetrap, and they are impatient to see it tested in the real world. "It would be an empty feeling," admits Harry Shorey, "to have done this work for so long just as a scientific game."

A major obstacle to the more rapid development and release of third-generation insecticides is the same law that enabled the Environmental Protection Agency to put many hazardous chemical poisons out of business—the Federal Environmental Pesticide Control Act of 1972—which is now inhibiting the progress of the safer insecticides it was presumably intended to encourage. An applicant for E.P.A. registration of any new insecticide must demonstrate not only that the formulation will do what it purports to, but also that it will not do anything else. It must be fed to rats, applied to the skin and eyes of rabbits and inhaled by two species of mammals, as presumptive proof that it will not hurt people. Generations of laboratory animals must survive exposure to it to demonstrate that it does not cause cancer or affect reproduction. The product's effects on wildlife and domestic animals must be tested on bobwhite quail and mallard ducks, rainbow trout and bluegill sunfish, cattle, earthworms and nontarget insects.

A complete E.P.A. registration application can run to several volumes of data covering several years of tests. According to companies that have paid it, the cost of answering the E.P.A.'s questions ranges from several hundred thousand dollars to several million. For the third-generation insecticides, the problem is not that they cannot meet the E.P.A. requirements for registration—it is generally agreed that the new materials are environmentally innocuous—but that they cannot justify the cost of filing the application.

A RISKY VENTURE

The manufacturer of a conventional insecticidal poison can anticipate recouping his development and testing costs from worldwide sales of his product for use against a number of pests. But the company contemplating production of a third-generation insecticide envisions a less inviting future. Because they are specifically designed to be effective only against their target pests and to be harmless to everything else, the third-generation insecticides require a new formulation, sometimes a new active compound, for each application. And each change in the insecticide requires a new E.P.A. registration, new tests, new costs. Finally, any entrepreneur willing to risk developing a hormone or pheromone insecticide may have difficulty patenting a product whose active ingredient is, after all, a natural substance.

As a result, while the agrichemical giants stick with their profitable poisons, the few smaller companies that have taken a chance on the third-generation insecticides complain that E.P.A. regulation is driving them out of the business. Zoecon, the only company founded specifically to develop and market third-generation insecticides, complains that it spent so much time and money getting E.P.A. registration for Altosid, its JH mosquito killer, that it is reluctant to do it again. Recently, according to one of its executives, when Zoecon had another JH formulation ready for use against whiteflies and aphids—both major crop pests—the company "chose not to generate enough data for an E.P.A. registration to use the product on food crops." Instead, Zoecon settled for a less profitable, but less expensive, permit to market the product for use on greenhouse ornamentals. In another case, Zoecon wanted to test a pheromone trap-out strategy similar to David Wood's. "The E.P.A. wanted so much data we dropped it," says a company spokesman.

Zoecon has now reportedly directed its program away from juvenile hormones and pheromones.

In an article in Science, Zoecon's president, Dr. Carl Djerassi, a professor of chemistry at Stanford, urges basic changes in the E.P.A.'s regulatory procedures. The E.P.A., as well as other regulatory agencies, says Djerassi, should be required to file "research impact statements" to make explicit the trade-offs between environmental safety and innovation. Djerassi also suggests that the E.P.A. relax its licensing requirements for experimental insecticides, and that E.P.A. tests be paid for by Federal loans, to be repaid when the new product is profitably marketed.

ARE REGULATIONS FAIR?

The entomologists themselves, convinced both of the need for strict environmental protection and of the safety and usefulness of their new insecticides, tend to take differing views of the E.P.A.'s requirements. David Wood's western pine beetle traps release pheromone in concentrations only slightly above the natural background level in remote forest areas. With the U.S. Forest Service sponsoring some of his research and running interference for his registration application in Washington, Wood hopes the E.P.A. will waive some of its expensive and time-consuming toxicology tests. Wood can live with the regulations.

Harry Shorey, on the other hand, releases relatively high concentrations of pink bollworm pheromone over very large areas where people cannot help but come in contact with it. According to the E.P.A. rules, Shorey could not even have field-tested his idea to find out if it would work. "The regulations are unreasonable," says Shorey, "particularly if you look at E.P.A.'s original charge to lessen the amount of toxic material in the environment." Some critics, like Shorey, maintain that the E.P.A. regulations, designed for the liabilities typical of the old poisons, are inappropriate for the new insecticides. It is relatively easy, for instance, to measure potentially dangerous residues of a conventional insecticide that is applied directly to plants at the rate of pounds per acre. But, J. Drew Horn of Conrel points out, it is quite a different matter to have to find for the E.P.A. residues of an evanescent pheromone that is atomized at the rate of only a few grams per acre per year. Other critics of the E.P.A.'s regulations contend that the third-generation compounds should enjoy a presumption of environmental innocence, on the ground that they are natural substances that have already been tested by eons of evolution. Dr. Thomas W. Brooks, Conrel's Venture Manager, goes further. E.P.A.'s reviews of the company's pheromone work "have almost completely missed the point," he says. "They are almost totally ignorant of the concept."

At the E.P.A., Dr. Martin H. Rogoff, the man in charge of pesticide regulation, replies that his agency is perfectly cognizant of the differences between conventional poisons and the new insecticides. Still, Rogoff points out, "being a natural substance doesn't necessarily mean that a hormone or a pheromone doesn't have toxic properties." Strychnine and arsenic, he notes, are "natural" substances. Despite the fact that "the quantities are minute and the hazards are small." Rosoff insists, any pesticide that is registered by the E.P.A. must have an "adequate demonstration" of what kind of hazard it represents. If the E.P.A. remains adamant, says Conrel's Brooks, "my company will not be able to justify continuing this line of research, nor will any other company. There's only a 50-50 chance of our surviving the present policy. The E.P.A. is jeopardizing the whole field of biologically rational insecticides."

Many insects are beneficial to man but none are as well-known as the honey bee. Here Virgil shows his patron Maecenas a bee-paradise.

Salmon Ranches, Lobster Farms

by Elisabeth Keiffer

An entrepreneur on the West Coast invested $50,000 in 1975 to install a large fish tank in his backyard and stock it with crabs flown in from Indonesia. He believed he had the beginnings of a lucrative new business known as aquaculture or, more simply, fish farming. What he didn't know, among a lot of other things, was that his backyard herd of crabs was fiercely cannibalistic. Penned together in the tank, the crabs began to eat one another. The man wound up with a $50,000 loss and one crab.

This unfortunate backyard businessman was one of hundreds of neophytes—giant corporations such as Union Carbide and Ralston Purina along with small-scale mom-and-pop operators—who have taken the plunge into fish farming in recent years. Certainly, in an ill-fed world, the notion of supplementing the sea's dwindling resources of marine protein is an appealing one. But what is really attracting venture capital these days is the prospect of soaring profits from America's insatiable market for such high-priced luxury seafoods as shrimp, salmon and lobster. Shrimp, for example, is so popular in the U.S. that, it is estimated, Americans alone could consume the world's supply by 1980.

"I'll bet in a week you could round up more people willing to risk money in an aquaculture enterprise than in almost any other kind of investment," says John Glude of the National Marine Fisheries Service, who is coordinator for the Federal Government's aquaculture program. "We get calls every day asking how to get started." As the parable of the West Coast crab entrepreneur suggests, getting started involves far more than merely stocking a pond with fish or crustaceans and reaping the harvest. For now, Glude advises would-be lobster farmers and salmon ranchers to wait until the technology is further advanced. The secret dream of every prospective aquaculturist—operating on a profitable assembly-line basis like chicken farmers—is still at least a decade away.

In fact, aquaculture is so new to the U.S. that only in recent years have the experts agreed how to spell it. Originally, it was "aquiculture"—taking the "i" from the science it most closely resembles, agriculture. When fish farming first began attracting interest in the U.S. in the 1960's, writers conjured up pastoral images of "plowing the sea," "fertilizing and weeding the ocean" and even "fencing off underwater farms electronically." Such visions

216

An imaginary weigh-in of a colossal shrimp after the roundup.

have not materialized and probably never will. The sea belongs to everyone; it is not up for sale or lease to aquacultural homesteaders. And the U.S. coastal and estuarine regions that would be ideally suited for aquaculture largely have been pre-empted by other uses such as leisure, oil drilling and, of course, conventional commercial fishing. Most future aquaculture, in fact, likely will be carried out in man-made saltwater ponds or tanks, possibly as far inland as Kansas City or Chicago.

Moreover, U.S. aquaculture's focus on luxury seafoods has forced it to begin largely from scratch. Though the practice of aquaculture is ancient in many parts of the world (some 2,500 years ago the Chinese scholar Fan Li wrote the

High-brow creatures such as lobsters and crabs are biologically complex and difficult to keep alive in captivity.

first surviving treatise on raising fish in confinement), it has concentrated on mass production of species that are considered unpalatable by the American market. For example, in Japan, where other forms of animal protein are in short supply, the average citizen consumers over 70 pounds of fish annually, compared with 12 pounds in the U.S. The Japanese raise millions of pounds of cheaply-fed, easily-cared-for fish like mullet, tilapia and milkfish. They simply introduce fast-breeding species into natural ponds where they feed themselves on algae, grow fat, reproduce and are harvested.

SEAFOOD SNOBS

By contrast, U.S. aquaculture must cater to a nation of seafood snobs who tend to disdain the lowly mullet and other easy-to-cultivate species. The problem is that high-brow creatures such as lobster, shrimp and crabs are far more complex biologically and far more difficult to keep alive in captivity. Worse, they are far more expensive to feed since they too prefer to dine on animal protein. To raise them profitably, American entrepreneurs have opted for "intensive" aquaculture which, unlike the Japanese pond culture, requires the development of complete mastery over the creature's entire life cycle. The faster it can be made to grow, the faster it can be sold, thereby returning a bigger, quicker profit.

Given science's scanty knowledge of the biology and ecology of many sea creatures, tinkering with the life cycle of a complex species like the salmon is no small order. For instance, how do you persuade salmon to reproduce in a tank? They have been driven by instinct to leave the sea and struggle upstream to their birthplace to spawn. How do you prevent cannibalistic young lobsters from dining off one another when they are crowded together? No commercial hatchery could afford the space to give each a private room for the 18 to 24 months it takes them to mature under optimum conditions. How do you speed up the growth rate of abalone? This high priced ($7 a pound in 1975) and fast-disappearing mollusk needs just the right seaweed diet for five to seven years to reach market size. How can a fish farmer be sure his crop, raised on artificial food, won't turn out

an unappetizing, unmarketable color? Salmon farmers discovered early on that, without some pigment-bearing shrimp shells in the ration, their salmon matured with white meat rather than the pink the public will pay for. In addition, there are engineering problems, such as providing a continuous supply of clean salt water of precisely the proper temperature to speed up growth—and then disposing of the water and the wastes it carries in a way that meets government standards. There are also legal problems—restrictions of siting, handling, marketing. And, assuming that all these problems can be solved, is it worth the effort? Will the farm-raised creature sell and turn a profit? Not long ago a computer calculated it would cost $21 a pound to cultivate lobster—some seven times the going retail rate for the traditional sea-harvested variety. This calculation helps explain the absence thus far of commercial lobster farms.

Nonetheless, enthusiasm for aquaculture is higher than ever. More than 100 U.S. businesses are involved in some aspect of marine aquaculture. Though many ventures are relatively modest, there are plenty of big names—among them, Union Carbide, Dow Chemical, Ralston Purina, Armour & Company, Sun Oil. Seaboard utility companies are also getting involved because they see a profitable use for their biggest by-product, the salt water that cools their generators. For encouragement, industry need look no further than the U.S. experience with farming freshwater fish, which are admittedly much easier to raise. Trout and catfish farms have been operating profitably in this country for some 30 years. This biggest single push for cultivation of marine species is being made by the Federal Government through N.O.A.A., the National Oceanic and Atmospheric Administration, in the Department of Commerce. Two of its agencies, the Office of Sea Grant, and the National Marine Fisheries Service, have put together a long-range aquaculture plan that called for a joint outlay of $5.5 million in 1976, and double that by 1979, to coordinate and fund research in universities and Government laboratories across the country. Their timetable is projected far into the future, with a heavy emphasis later on genetic research. Aquaculture's ultimate triumph would be to achieve, through selective

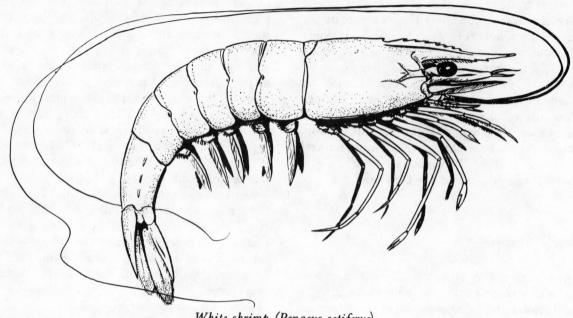

White shrimp (Penaeus setiferus).

breeding, generations of marine animals that are fast-growing, plump-meated, disease-resistant and noncannibalistic.

SHRIMP FARMS

Current commercial interest focuses on salmon, oysters and shrimp. The U.S. consumes shrimp at a rate of over one million pounds a day, 65 percent of it imported. The only commercial-scale shrimp farm is operated by Marifarms Inc., a Florida company that has spent nine years and more than $10 million and is only now emerging from the red. Only a fluke (the meteorological kind) kept it from turning a profit in 1976. Just as the company was ready to harvest 1½ million pounds at its Panama City farm, Hurricane Eloise swept the entire crop into the bay, where it provided rich pickings for shrimp trawlers from miles around. But Marifarms' president, John R. Cheshire, a chemical engineer formerly with Du Pont, is not discouraged. "Do you realize it took 20 years to scale up nylon?" he asks. "I think we can really crack this field open." When that happens, he says, consumers will enjoy a product far superior in quality to shrimp caught at sea and brought into port days or weeks later.

Unlike Marifarms' big operation, Government-funded shrimp farms are small pilot plants. Their aim is not immediate profit but devising a model system efficient enough to tempt even the most cautious investor to "scale it up." Most scientists believe that the capability of inducing captive saltwater shrimp to spawn reliably will be essential to such a system. Without reliable spawning, stock has to be gotten from pregnant wild females, an undependable and expensive process. "It's like having to run out to the prairie for a cow every time you need calves," says Dr. Fred S. Conte, a marine biologist at Texas A. & M. University's pilot farm, who has devoted the past few years to this problem. A further advantage of having captive seed stock, he points out, is the possibility it offers for selective breeding programs. Texas A. & M. has had a N.O.A.A. Sea Grant since 1968 to develop shrimp culture in cooperation with the Texas Agriculture Extension Service. The Ralston Purina Company and the Central Power and Light Company also give it financial support. Not long ago, project director Dr. Jack Parker invited me to fly over its two pilot farms to see how shrimp farming may be practiced in

A male breeder shrimp.

that state a few years hence—as routinely as rice or soybean farming.

Along the Gulf coast lie more than two million acres of underutilized lowlands. They rent for as little as $3 per acre or sell for $300. Some day they may be covered with neat rows of rectangular man-made ponds, fed by water pumped from Galveston Bay and jumping with shrimp. Dr. Parker visualizes an operation modeled closely on American agricultural methods, with feeding, harvesting, processing and packaging handled by automation. The fixed costs of starting up a 500-acre farm along these lines have been estimated by the university at $1.5 million, with annual running expenses

projected at the same figure. If everything worked perfectly on this hypothetical farm, which, Dr. Parker concedes, rarely happens, it could produce 2.5 million pounds of shrimp annually with a market value of $2.8 million. Six or seven people, he says, should be able to handle the entire operation, and he predicts that in six or seven years they will be doing it.

BREEDING IN CAPTIVITY

Why things rarely do work perfectly is illustrated by Dr. Parker's description of a devilishly complex process—the hatching of baby shrimp in captivity. The eggs must first be taken from females in the wild, then hatched in special laboratory tanks. A new larva is one of perhaps half a million barely visible specks suspended in the water. Easily crushed, unable to swim or feed itself, it depends on artificially generated turbulence in the water to keep it from sinking and to make certain that food—microscopic phytoplankton—collides with its mouth. In the first two weeks of life it goes through 12 enormously complex larval transformations to become at last a recognizible but minute "baby," some 20 millimeters long and weighing about 0.02 grams. As the baby grows, its food preferences change from vegetable to animal, and in its first postlarval days it greedily devours its brethren if it is not quickly sated with tastier food. In another two weeks, the surviving babies are ready to leave the hatching tanks and go into "grow-out" ponds—quarter-acre cement rectangles that, when viewed from our plane 500 feet up, looked like 18 dormitory cots flanking a central aisle. Each held 10,000 to 50,000 quarter-inch-long shrimp. With luck, the shrimp that were now feeding on fish food formulated and supplied by Ralston Purina would reach market size in 110 to 114 days. Then the ponds would be drained and the crops harvested.

The site of this prototype shrimp farm near Corpus Christi had been chosen with great care. Nearby is a generating plant of the Central Power and Light Company, which takes its water from Laguna Madre, a bay on the Gulf of Mexico. Scientists wanted to find out whether shrimp growth would be affected by the higher-than-average salinity of Laguna Madre water and by the warm temperature it reaches when passing through the plant's power generators. The water, they learned, speeds up shrimp growth. Now they are experimenting in hopes of finding an ideal temperature and degree of salinity.

On another front, Dr. Parker and his associates have achieved what appears to be an exciting breakthrough in shrimp biology. Last year, for the first time, 100 shrimp that had been isolated not only reached sexual maturity in captivity but mated there and produced fertilized eggs. Morever, Dr. Conte had a hunch that even more shrimp would mature sexually if allowed to winter over in the ponds. To keep the water at the proper temperature, he covered them with a "blanket" of solar-energy collectors. Last spring his hunch proved right: There were 1,400 sexually mature shrimp. But 100 percent success in breeding is not assured yet. Even slight interferences with the natural process can totally disrupt it. When female shrimp are disturbed, for instance, they will reabsorb their eggs before they can be fertilized. "But we're pretty sure we're on the right track now," Dr. Conte says. The scientists believe it is only a matter of time before reproduction in captivity can be guaranteed. When that day comes, they say, Texas will be ready for commercial shrimp farming.

Salmon and oysters, like shrimp, have a high market value and, because a primitive form of aquaculture has been practiced with them in Government hatcheries for over a century, their requirements are better understood than those of many sea creatures. As a matter of fact, the U.S. now leads the world in oyster harvest—and 40 percent of the oysters are grown on farms. The problem with oysters is the market. Though once a staple of the working-class American's diet, oysters became so expensively scarce that several generations of consumers grew up without ever tasting an oyster. The trick now is to persuade Americans that they ought at least to try one.

COHO SALMON

The Atlantic salmon, considered the king of fish by sportsmen and gourmets, disappeared from New England's rivers years ago when mills and then utility companies blocked the way to age-old spawning grounds. But two competing entrepreneurs (one of them a former Boston stockbroker, the other a marine biologist) have come up with a handsome replacement—the coho salmon, a Pacific species whose eggs are flown East, raised to small-pan size in 18 months and marketed mainly to restaurants along the Eastern Seaboard. To raise their coho salmon, both companies have chosen sites in Maine that offer highly favorable local circumstances. At Harborside, Maine Sea Farms utilizes a 320-foot-deep pit left by an old intertidal copper

Checking strings of oyster shells.

mine. Flooded and returned to tidal exchange, the pit made a perfect nursery: it has just the proper temperature and salinity and much of the feed needed for the 300,000 cohos being raised there in net pens. Meanwhile, Maine Salmon Farms of Wiscasset grows its cohos in pens in the Sheepscot River, taking advantage of the warm-water discharge from a Central Maine Power Company plant to speed up growth. Both operations have been able to hang on despite the disasters, man-made and natural, that seem to confront every pioneer aquaculture venture, and both are confident that they will be making money soon.

An even more innovative form of salmon farming, being tried on the West Coast, is ocean ranching—raising eggs to migrating age in coastal rearing facilities, releasing the young to the sea and harvesting them when they come back two to four years later. The beauty of this method is that it eliminates a major cost—feeding and maintaining the fish for much of their lives. Weyerhaeuser Company thinks the

real growth potential for fish farming lies in this direction, and the Federal Government's N.O.A.A. is equally enthusiastic. "But I'm not encouraging people to jump into the business too fast," John Glude says. "There'll be a shortage of seed stock till the first crops are harvested, and there are legal problems to be worked out in some states."

THE DIFFICULT LOBSTER

The Hope diamond of all these Tiffany seafoods is, of course, lobster. But, by N.O.A.A.'s estimates, big-scale commercial farming appears at least 15 years away. Even though *Homarus americanus* is now relatively well understood and can be raised in captivity, the process is still too far from profitability, principally because an inexpensive food that will make lobsters grow fast remains to be found. Meanwhile, scientists are looking into other

223

A string of oysters after suspension from a raft for 18 months in a Cape Cod experiment.

Feeding time for salmon being reared in captivity.

approaches to increasing lobster production. San Diego State University biologists have imported *Homarus* from Massachusetts and are test-raising them in shore-based corrals heated by the San Diego Gas and Electric Company's power-plant effluents. At the University of Rhode Island on Narragansett Bay, diver-scientists have placed cinderblock "habitats" in strategic locations on the ocean floor to lure the shy, shelter-demanding crustaceans to new areas. Pilot studies were so promising—all but four of the original 195 apartments attracted tenants within two months—that the experiment is being considerably enlarged.

Government-backed research into farming lobster and other luxury seafoods obviously raises questions about priorities in a world hungry for cheap sources of protein. Why, for example, doesn't the U.S. encourage the multiplication of that easy-to-raise creature tilapia, the fish with which Christ fed the multitudes.

Lobster and shrimp are not only too expensive for most of the people in the world, their cultivation actually results in less protein since, like cattle-raising, it requires large amounts of protein-based feed. Indeed, Government experts agree that, perhaps a decade from now, American aquaculture may need to take a totally different direction—mass-producing low-cost, fast-growing, plant-eating fish for use in new, high-protein processed foods. Meanwhile, they say, the Federal Government wants to encourage growth of a healthy aquaculture industry that can generate employment, supplement the sea's resources and, if necessary, eventually switch to mass production of low-cost marine protein. To stimulate that growth, substantial private investment is needed. Hence the Government is plugging the glamour stocks such as lobster and shrimp. After all, how many investors would be keen on taking a flyer in mullet this year?

Scientists Brave Bears and Winter in Hunt for a Sleep Hormone

by Lawrence K. Altman

In open-air cages on a farm at the Mayo Clinic in Rochester, Minnesota, two black bears have gained more than 100 pounds each after feasting around-the-clock for several weeks preparing to go into a long winter sleep. The bears will then be moved into dens where, from time to time, a team of Mayo scientists will join them for stretches lasting as long as 36 hours and in temperatures as low as minus 30 degrees. The scientists will be drawing a series of samples of bear blood.

This will be the eighth winter that Dr. Ralph A. Nelson's team here has tried to unlock one of nature's greatest secrets—how, even when amply supplied with fat, the bear can sleep for up to five months, burning 4,000 calories each day, and yet not once eat, drink, urinate or deficate. The scientists are seeking a hormone that they suspect controls the bear's winter sleep pattern. Discovery of such a hormone, they believe, might offer new ways to treat human diseases such as kidney failure, sleep disorders, obesity and starvation, and open new avenues of research into many other conditions.

Data from the bear studies have led the scientists to devise a low-protein, low-fluid diet for patients with kidney failure. Studies on Mayo Clinic pa-tients without kidneys have shown that such patients can go 10 days instead of the usual three before they need another treatment with an artificial kidney machine.

The study team includes physicians, veterinarians, medical students and technicians. In working with the bears, they brave not only Minnesota's harsh winters but also the dangers involved in arousing and sticking needles into the 400-pound animals to inject radioactive substances, then to withdraw blood samples periodically. By testing the blood samples in a laboratory to follow the radioactive decay pattern, the scientists learn how the bears can use the stored foodstuffs—fats, proteins and sugars—so precisely that they do not need extra calories during the period of "hibernation" from December to March.

Although wild bears sleep up to five months in the winter, they are not true hibernators like woodchucks, ground squirrels and many reptiles. The bear's temperature drops by just four degrees, whereas those of true hibernators plummet more than 60 degrees. Unlike many small hibernators that sleep so soundly they can be picked up and tossed at will, a bear in winter sleep will arise at the slightest noise, charge at a visitor and, if suffi-

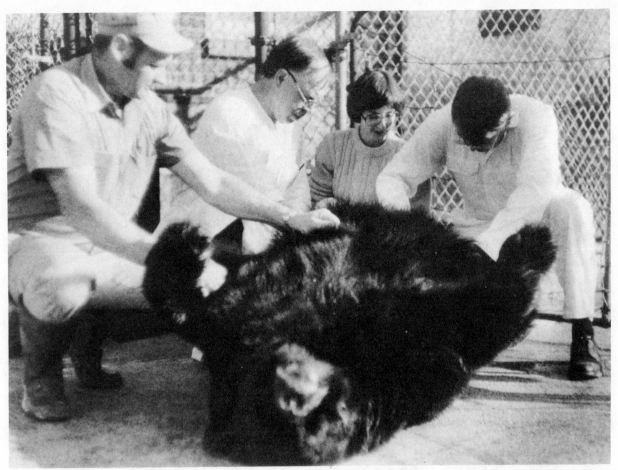

Dr. Ralph A. Nelson, right and members of his scientific team taking blood from one of the bears to use for comparison with later samples to be taken during the hibernation period.

ciently concerned about the disturbance, even move its den.

BEAR SPAT WHEN DISTURBED

One day, when Dianne Wellik, a research assistant, and a visitor approached the cage with a syringe and needle attached to a long pole, the bear stood on its hind legs, grabbed the mesh with its front paws and spat. Bears can be subdued for the experiments with injections of a combination of muscle-relaxing and anesthetic drugs. When the drugs took effect, the team carried the bear from the cage into the open, where it was placed on its back.

Mrs. Wellik ran her fingers over the bear's lower abdomen until she felt a pulse that guided her to the femoral vein. Then she took a large syringe

and needle and withdrew samples of blood. A few minutes later, after the tests were completed, the bear was carried back to the cage and placed in a sitting position to help prevent it from coughing up its stomach contents and developing pneumonia.

The bear experiments began with a casual remark made to Dr. Nelson eight years ago by a doctor who had just returned from a medical meeting where he heard about the bear's winter sleeping characteristics. Dr. Nelson, a physician who also has a Ph.D. degree in physiology and who specializes in nutrition, was puzzled: How could the bears survive while accumulating toxic waste products? Dr. Nelson, who enjoys telling stories about his bears as much as doing the research, laughed as he recalled how the first bear arrived at the Mayo Clinic from the Upper Peninsula in Michigan on the back seat of a Volkswagen. The bear had been calmed for the 300-mile trip with

Black bears feeding in their cages on farm at the Mayo Clinic in Rochester, Minnesota, before they begin hibernation under observation.

injections of a muscle-relaxing drug.

"Everyone who works with bears gains an immediate respect and love for these animals," Dr. Nelson said. "They go without food and water, yet the females can incubate fetuses and then nurse cubs during winter sleep. Other animals can't find the bear because by producing no urine or feces he leaves no odor."

When the team first began testing the bears, Dr. Nelson said, Dr. Paul E. Zollman, a veterinarian, would put a syringe containing an anesthetic on the end of a stick, walk in, back the bear into a corner of the cage, stick him and rush out. "We couldn't ask Paul to risk his life all the time," he added, "so we used our ingenuity, got a longer pole and learned how to inject the bear from outside the cage."

QUIET SPOT IS NEEDED

Some bears fail to hibernate in the wild, and the Mayo team was told that the bears would have even more difficulty hibernating in captivity. The researchers learned that the bears would hibernate only if they denned in a quiet spot. The Mayo bears den on the farm in what was a root cellar.

When the studies are done in the hibernating season, Mrs. Wellik keeps warm by carefully huddling against the anesthetized bear's fur and by wearing battery-powered socks. But her hands are numbed by the cold because she cannot use gloves when she guides the syringe and needle into the bear's veins.

The researchers work strictly with private money and donate personal lecture fees to futher

Dr. Ralph A. Nelson

their studies. Because the logistics make it too difficult to work with more than three bears at a time, the research has proceeded more slowly than it might have if other animals were used.

ADVANTAGES TO LARGE ANIMALS

"People ask why we study such a large, dangerous, mean animal and not the smaller hibernating woodchuck," Dr. Nelson said. "But there are advantages working with large animals. With a small

animal it is difficult to do tests on small amounts of blood and tissues. Some studies we do on the bear would be impossible to do on a smaller animal because we would bleed it to death taking the samples we need," he said. "The bear is so large your chance of isolating substances is much better."

The Mayo team usually does an experiment one winter, repeats it in the nonhibernating season for comparative purposes, and then confirms results the next winter. Each year a new experiment is

229

added. The team plans well ahead because a missed opportunity means waiting another year. Interpretation of the data collected reflects hours of thought and observation. Dr. Nelson said he was taught "that doing research is like receiving a cablegram—from just a few words you have to figure out what's going on back there," By such insights, he said, "we know what happens, how it happens, but we don't know what switches on the bears." Over the years, Dr. Nelson's team has learned that the bear's reaction to hibernation resembles the human's in starvation "except that the bear does it perfectly." Starving or bedridden humans burn up fat but they also lose protein as muscles deteriorate.

WILD BEAR BECOMES RAVENOUS

But the bear tolerates winter sleep very well. For about one month before going into it, the wild bear becomes ravenous, eating 20 out of every 24 hours, increasing his daily caloric intake to 20,000 from about 7,000, and gaining more than 100 pounds. Then, once the energy input stops and sleep begins, the bear's biochemical reactions become delicately balanced. The animal makes just enough water from its fat stores to stay hydrated. Normally the bear, like other animals, manufactures and breaks down proteins at a constant rate in a process called "protein turnover."

However, the researchers found that protein

One of the bears in a man-made den which will be its winter quarters.

Caretakers occasionally enter the cages to feed the bears.

turnover speeds up five times during hibernation without increasing the amount of protein in the body at any one time. This conservation program provides the bear with as much protein at the end as at the beginning of winter sleep.

FAT LEVEL RISES

The process is so efficient that the bear does not form excess amounts of urea, the waste product of protein breakdown, which is excreted in urine and

231

which becomes toxic when large quantities build up as a result of damaged kidneys. In hibernation, the bear's two kidneys produce just a trickle of urine and that is reabsorbed into the blood through the bladder wall.

The bear's adjustment to hibernation is so sophisticated that results of most standard laboratory tests of blood samples from the nonhibernating and hibernating periods are the same. The only exception is that cholesterol and other fats rise during hibernation. When the bear wakes up in the spring after a three month fast, it still is not hungry. "We can't get him to eat for two weeks even when we serve him food," Dr Nelson said.

The researchers have found that bears that cannot hibernate starve to death just like a human or other animal. And because the researchers have tried unsuccessfully to get bears to hibernate in the summer by putting them in dark, cold rooms mimicking winter conditions, they suspect the phenomenon is controlled by a hormone that the brain's hypothalamus, which regulates many basic body functions, such as temperature, makes each fall and winter.

Dr. Nelson speculates that discovery of a hormone controlling winter sleep might help patients with kidney disease as insulin helps people with diabetes. He also theorizes that because the bears lose their appetite in the face of food, such a hormone might prove helpful in treating obese people. At the same time, it might help prevent malnutrition among the starving peoples of the world. "The bear goes through both extremes—starvation and obesity—each year, but the bear treats obesity as no human can," Dr. Nelson said.

Nobel Laureate Seeking to Unravel the Shape of an Enzyme

by Malcolm W. Browne

Some revolutionary tools for probing life's secrets are taking shape under the hand of William Nunn Lipscomb Jr., winner of the 1977 Nobel Prize in chemistry. Dr. Lipscomb's new work, in the view of many colleagues worldwide, is so significant that it could well win him a second Nobel Prize. A tall, affable Kentuckian, he smiles self-deprecatingly at mention of Nobel Prizes but acknowledges that his current research is the most important he has done.

The three decades of work that led to Dr. Lipscomb's Nobel Prize were mainly with a class of chemical compounds called boranes, molecules made of boron and hydrogen. His current work is chiefly with a regulatory enzyme called aspartate transcarbamylase or ATCase, one of the key triggers of cell division and growth in all living things. Biochemists, biologists and doctors are watching with great interest, since the work of the Gibbs Laboratory of Harvard University, over which Dr. Lipscomb presides, promises to shed light on such vitally practical matters as cancer cell division. Among scientists who have expressed special praise for Dr. Lipscomb's current work is one of his former students, Dr. Raoul Hoffman, a Cornell University chemist. "His work is particularly remarkable now, not only because of its intrinsic value but because it is so unlike his earlier work in boranes," Dr. Hoffman said. "It is, in my view, as likely to win him a Nobel Prize as was the borane work."

Paradoxically, Dr. Lipscomb is a physical chemist, not a biologist or even an organic chemist. The ideas with which he works have to do with geometric shapes and transformations, quantum mechanics and relativity theory, seemingly more in the realm of lifeless abstractions than microbes or living tissue. Dr. Lipscomb, 57 years old, has contended through most of his career that biology cannot be fully understood without explanations from mathematics and physics, and his views now seen fully vindicated. "The old boundaries between the different branches of science are breaking down fast," he said in an interview. "For many years, the biochemists have been at the fore with their dramatic discoveries in DNA research and so forth, but the major breakthroughs from here on, I believe, are likely to come from the fundamental branches, physics and physical chemistry."

SHAPE FOUND IMPORTANT

Dr. Lipscomb had been impressed by the knowledge gained from earlier studies that the

Dr. William Nunn Lipscomb Jr., with a molecular model in his laboratory.

way atoms interact within a molecule depends not only on the classical ideas of electron bonding, involving the interchange of electrons, but also on the respective positions of the various atoms in three-dimensional space. As the physical shapes of certain molecules are warped or distorted, it was found, the electronic bonds between their constituent atoms may be radically changed, changing the chemical behavior of the whole molecule.

Such phenomena could explain some of the mysteries of the behavior of complex molecules in living systems, Dr. Lipscomb believed, and he decided a logical start could be made with ATCase. Like the hundreds of other enzymes, the substance catalyzes chemical reactions without taking part in them. The exact physical shape of the molecule could explain how it triggered cell division, he hypothesized.

The task of devising a three-dimensional map of the ATCase molecule now occupies most of Dr. Lipscomb's research time. (He also teaches full-time for four months a year and travels to participate in many symposia in other months.) Superficially, there seems to be an enormous gulf between the lifeless, exotic boranes and such a huge, complicated molecule as ATCase, which is essential to life.

A LOGICAL OUTGROWTH

But for Dr. Lipscomb, enzyme research is a logical outgrowth of his borane work, which is also continuing, because the computer techniques he developed to determine the latter's structure can be extended to investigations of the former, which is far more complex. "I periodically decide to give up borane research, imagining that there is little more to discover," he said. "But I have never abandoned boranes for more than six months, because some new idea always brings me back, making me realize how much more there is to be learned about chemistry in general from boranes than I had previously imagined."

For many years, fellow chemists regarded Dr. Lipscomb as mildly eccentric, not merely because he affects bright checked shirts and string ties, or because he has a passion for chamber music and great skill with the clarinet. For an aspiring young chemist in the 1940's, few subjects could seem so esoteric and so unlikely to result in major scientific discovery (or practical applications) as boron chemistry. Boron research at the time seemed to most chemists an utter waste of time. But Dr. Lipscomb had become intrigued by the subject and stimulated by the thinking of his teacher and mentor, the Nobel laureate, Linus Pauling. "The more I thought about boranes," Dr. Lipscomb said, "the more I came to realize the importance of molecular relationships between structure and function." Thus, Dr. Lipscomb explained, the same atoms arranged in a three-dimensional molecule may behave quite differently when twisted or pushed out of shape with respect to each other. Such proved to be the case with the boranes, strange molecules shaped like cages, some having 20 faces or more.

Dr. Lipscomb, and the dozen or so doctoral and post-doctoral students who perform the actual experimental work for him, determined the shape of many of the boranes using a device called the X-ray diffractometer, a table-top X-ray machine for looking at tiny crystals. But before many of the new boranes were actually discovered, Dr. Lipscomb had predicted their existence with the aid of higher mathematics and a children's construction toy called "D-Stix," from which complicated models took shape. Such tools had been refined to a high degree in his borane work when Dr. Lipscomb decided to take on ATCase, a winding, convoluted tangle of amino acids and related substances. It had long been established that the enzyme was related to cell division, but an understanding of how the enzyme's action is triggered has defied biochemists.

ANSWER IN STRUCTURE

Dr. Lipscomb reasoned that since chemical function always depends on structure, an exact knowledge of the shape of the ATCase molecule would ultimately answer the question. A key part of that question is how a chemical signal impinging on one of the extremities of the molecule is transmitted over the relatively huge distance of 40 Angstrom units (40 ten-billionths of a meter) to its "active center." The "active center," the part of the molecule that causes it to trigger division in other cells, is in a cavity indented in the center of its structure.

To learn much about the matter requires de-

Dr. Lipscomb is seen here working on a model for an enzyme molecule.

termining the spatial relationship of every atom in the molecule to every other atom in the vast structure. The magnitude of the task, which is likely to continue for several more years, is evident from the molecular weight of ATCase: more than 300,000. (The molecular weight of the simplest molecule, hydrogen, is about 2.) It is slow work, but improvements in the tools have helped. The circular patterns of dots that result from Dr. Lipscomb's X-ray diffraction photographs are now scanned automatically and the results translated into computer language.

COMPUTERS DRAW MAPS

Computers have been taught to draw topographical maps using the data, maps that look like those made by geologists to depict mountains and valleys. The contours derived from the X-ray data, however, mark the positions and spatial relationships of atoms and clusters of atoms inside a molecule. The resolution, or focus, of the pictures thus obtained has been described by Dr. Lipscomb's admirers as a triumph of scientific technique. Similar techniques enhanced by Dr. Lipscomb and his co-workers have resulted in a wholly new way of looking at chemical bonding, with the realization that a single electron can bind as many as three atoms together within a molecule, not just two.

"To give you an idea of how radically chemistry has changed over the years," Dr. Lipscomb said, "I can say that I spend practically no time in a laboratory. At the Gibbs Laboratory we do no chemical analysis and practically no synthesis. But by contrast, I probably use more time on Harvard's computers and other computers available to us than any individual—huge hunks of computer time. It's a far cry from my own antecedents. I became interested in chemistry when my mother gave me an A.C. Gilbert chemistry set. But today the real research is in ideas—even intuition— expressed, usually, in the special languages with which mathematicians and computers communicate."

Dr. Lipscomb communicates many of his ideas to colleagues and students with visual aids, models and three-dimensional photographs and drawings that must be viewed with stereoscopic devices. One of his key assistants is a former Radcliffe College tennis coach, Jean Evans, an artist who works at the Gibbs Laboratory drawing the diagrams and sketches needed to make the chemist's ideas easily intelligible. Dr. Lipscomb, on the other hand, has become an ardent tennis player.

Held in great affection by colleagues, who invariably address him as "Colonel" in recognition of his Kentucky origins and honorary colonel's rank, Dr. Lipscomb teaches that the creative process should be kept apart from purely practical endeavor. "For me," he said, "the creative process, first of all, requires a good nine hours of sleep a night. Second, it must not be pushed too hard by the need to produce practical applications. Without fundamental, basic research, we quickly run out of the discoveries that give the engineers and therapists and other practical people the material to advance. In some ways, our crash approach to finding a cure for cancer has been mistaken, I believe," he said. "The funding should provide more emphasis on basic research. As for cancer itself, I don't believe a single cure will ever be found, because it is not one disease but hundreds of different ones all described by the same name. But through basic research we can chip away at each disease individually. And, which for me is equally important, we can increase man's fundamental knowledge of the universe of which he is part."

Scientists Discover New Form of Life That Predates Higher Organisms

by Richard D. Lyons

Scientists studying the evolution of primitive organisms reported in November 1977 the existence of a separate form of life that is hard to find in nature. They described it as a "third kingdom" of living material, composed of ancestral cells that abhor oxygen, digest carbon dioxide and produce methane.

The research group working at the University of Illinois reported that this third form of life on earth was genetically distinct from the higher organisms that evolved from it—bacteria and, finally, the plant and animal world. Bacteria, with their own distinct form of cells, are more primitive than plant and animal life, which have vastly more complicated cellular structures.

Believed to have evolved 3.5 billion to 4 billion years ago, these organisms have yet to be named but are being referred to informally as either archaebacteria or methanogens. Before this report, the oldest form of life, bacteria, was believed to have evolved about 3.4 billion years ago. "We have shown that they are genetically distinct from the higher organisms," said Dr. Carl R. Woese, the leader of the group investigating the evolution of microorganisms.

The genetic tracking efforts of the scientific group, which spanned five years, were made public by two of the Federal agencies that supported the research, the National Aeronautics and Space Administration and the National Science Foundation. The work is described in detail in the October and November 1977 issues of the Proceedings of the National Academy of Sciences. Asked for their evaluation of the results of the team at the University of Illinois, two other scientists familiar with the genetics of microbiology described the report as "important" and "exciting," adding that it would further what is known of the basic processes of evolution.

Dr. Woese and his colleagues conclude that before the emergence on the earth of bacteria, usually regarded as the simplest form of life as we know it, at least one and perhaps several earlier forms of primitive organisms had evolved from the primordial ooze that developed after the crust of the earth had been solidified from a gaseous cloud.

Dr. Woese, whose name is pronounced "woes," said in an interview that the practical value of the research probably was nil. But he added that, if the

Dr. Carl R. Woese, leader of research team investigating the evolution of microorganism, in his office at the University of Illinois.

efforts of his group were confirmed by other researchers, the findings would enhance man's knowledge of his genetic heritage and perhaps explain some of the mysteries of evolution and puzzles of the solar system. One flight of fancy advanced by Dr. Woese, with an accompanying smile, is that the presence of this class of organisms might explain why life evolved here and not on the earth's sister planet, Venus.

The rationale goes as follows: Clouds of carbon dioxide originally enveloped both planets, but methanogens developed on earth and digested much of the cloud and in turn produced the hydrocarbons that developed into higher forms of life. But on Venus, according to this line of speculation, the lack of methanogens allowed the carbon

dioxide to accumulate to the point that the so-called "greenhouse effect" took over the Venusian surface, in turn making it too hot for life to evolve.

Dr. Woese, a slightly built biophysicist who has an unruly shock of graying hair, expounded on the research of his group in a three-hour interview in his sparsely furnished office here. Nearby rooms are filled with such gadgets as electron microscopes and X-ray machines that are the basic tools used in deciphering the genetics of microorganisms. "For years I've wanted to understand how life evolved," he said, "and five years ago my colleagues and I set about looking into the genealogy of organisms."

Dr. Woese, who is 49 years old, said that only in the last 10 years has it been feasible to explore the

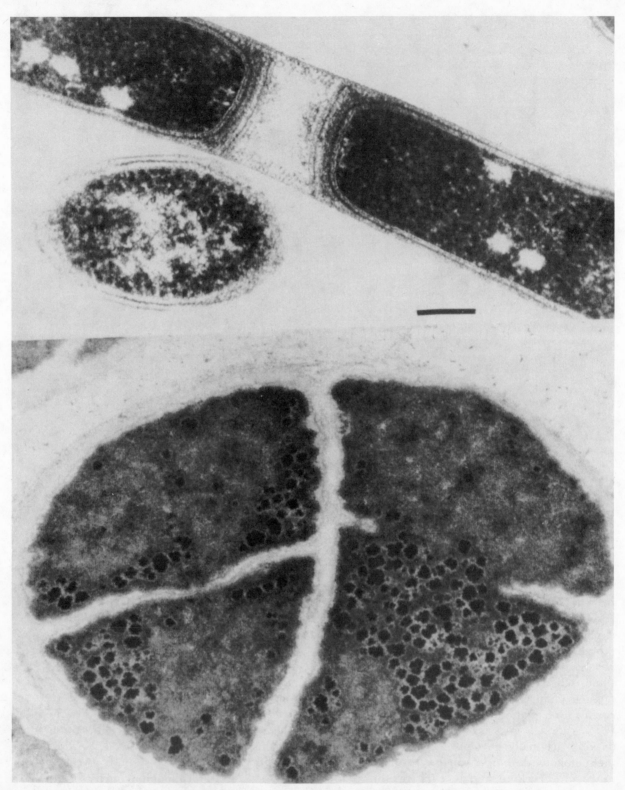

This photograph shows the newly discovered microorganism: top, a chain of two organisms, each one-thousandth-of-a millimeter long; center, a cross section of the chain; bottom, an organism dividing into four cells.

genetics of such rudimentary organisms. Elaborating, he cited the existence of only an elementary knowledge of molecular genetics a decade ago, the development of more powerful electron microscopes, and the discovery of more sophisticated techniques for examining the molecular structures of microorganisms.

At first the group examined DNA of bacteria, that is, the deoxyribonucleic acid molecules that contain the coded information needed for the function and development of the cell. The team also studied bacterial ribosomal RNA, the ribonucleic acid that is a major constituent of the ribosomes. These are the units within cells where the messages from the genes are received and read in order to make the appropriate proteins. The ribosomal RNA's are believed to be extremely old and represent parts of the ancestral replicating systems of both primitive and advanced organisms.

Examining the parts of either an animal cell or a plant cell is relatively easy as compared with a bacterial cell, which is perhaps 1,000 times smaller. Also, the bacterial cell does not have a clearly defined structure that the higher forms of life possess.

According to Dr. Woese, the early research on the evolution of microorganisms focused on their structural differences, rather than their genetic differences. By examining ever simpler forms of bacteria, the University of Illinois scientists arrived at what then were believed to be the simplest forms, which the scientists have now found not to be bacteria at all.

"The methanogens themselves are not new to science," Dr. Woese said. He noted that 10 different forms had been examined in the course of the research here and that their total number was unknown because "scientists have just begun to isolate them in earnest and there could be millions of them." To be examined, the methanogens must be cultured under extremely difficult conditions since they will not exist in the presence of oxygen.

Various forms of methanogens have been found in mud at the bottoms of San Francisco Bay and the Black Sea, in deposits in Carioco Bay off the coast of Venezuela, and in deep, hot spring waters such as those at Yellowstone National Park. They generally are found in what are called anaerobic niches, or areas free of the presence of oxygen, which are relatively uncommon on the earth's surface.

The technique used here cultured the methanogens in the presence of radioactive phosphorus, which in turn made the RNA radioactive. The radioactive RNA then was separated from the genes through the use of acrylamide gel electrophoresis. The RNA then was digested with enzymes into smaller pieces and their molecular sequences, or messages as they are called, were compared with the RNA messages of either higher or lower organisms.

"Somewhere along the line in evolution a mistake is made and a mutation results," Dr. Woese said, adding that by studying these mutations it was possible to compare the ages of different RNA's. By deciphering the mutations of the genetic material, the scientists were able to identify methanogens as being distinctly different from bacteria.

Dr. Woese credited the name "methanogen" as having been coined by a colleague on the project, Dr. Ralph S. Wolfe, a professor of microbiology. Other collaborators included Linda J. Magrum, a research assistant; William E. Balch, a graduate student, and Dr. George E. Fox, the senior author of the paper in the October 1977 issue of the proceedings, who is now an assistant professor of biophysical sciences at the University of Houston.

Asked for comment about the University of Illinois work, Dr. Sol Spiegelman, now a professor of genetics at Columbia University, said that "the research results look O.K. Dr. Woese is a substantial scientist of international reputation who has contributed a number of ingenious ideas to science."

Dr. Cyril Ponnamperuma, director of the laboratory of chemical evolution at the University of Maryland who has reported on extraterrestrial organic molecules, described the work as "very exciting, even fantastic. It fits into the general idea of evolution under nonoxygen conditions."

Inbred Mice

by Jane E. Brody

As Anna Stanley went about her chores one afternoon at the Jackson Laboratory in Bar Harbor, Maine—cleaning the mouse cages, replenishing the food and water—she spotted a deformed baby mouse in a 5-day-old litter of laboratory mice. The tiny creature had been born with no hind legs and a deformed front leg.

The discovery caused quite a stir in the animal room because Jackson Lab, which has just celebrated its 50th anniversary, is the repository for 70 percent of the world's known mouse mutants and the largest collection of inbred mice. Mutations that spontaneously occurred at the lab have given the world's researchers genetic models of such human illnesses as diabetes, obesity, muscular dystrophy, inherited anemias, leukemia and breast cancer.

With these animals, scientists can perform experiments that would be impossible or unethical to do in human beings. And, by incorporating the mutations into inbred strains of mice that are genetic carbon copies of one another, scientists can study the effects of single genes unconfused by the enormous genetic variability of ordinary animals and people.

The discovery wasn't the first anomalous mouse Anna Stanley, who is 59, has found during her 12 years at Jackson Laboratory, where three million descendants of the common house mouse are born each year. In years past, she has found among these inbred mice, which represent the world's most valuable tool for biological and medical research, a hairless dwarf, a mouse that danced in tight circles and one with a corkscrew tail. Jackson scientists examine each oddball mouse discovered in a new litter and decide whether to attempt to breed it in order to determine if the defect is caused by a genetic mutation. In the case of the hairless dwarf and the dancing mouse, both oddities were shown to be caused by genetic defects and the descendants of these mice—bred to pass the defect from generation to generation—are now being studied for hormonal and balance disturbances that may reveal new information about similar disturbances in humans.

If Jackson Laboratory follows its normal Tuesday routine, the mouse Anna Stanley found will be transferred to the laboratory's mutant mouse

Inherited anemias, and the potential of marrow transplants to cure some forms of anemia as well as some forms of cancer, are the specialty of Dr. Seldon Bernstein.

A mouse mother from DBA inbred strain of mice watches over her young. The DBA inbred strain, the oldest in the world, has been inbred since 1909.

Anna Stanley goes about her work caring for and checking thousands of mice each week. Animal caretakers are trained to be alert for possible new mutations.

are genetically different, it's harder to tell what caused what.

The late Dr. Clarence C. Little, who founded the laboratory in 1929, developed the first strain of inbred mice, called DBA. These mice are genetically prone to noise-induced epileptic seizures and breast cancers. He needed the genetically identical animals to carry out his cancer studies, in which he transplanted tumors from one animal to another. The transplants would be rejected as foreign unless the animals were identical twins, so he created in effect an endless family of such twins. Dr. Little, who initially had to "sell" the scientific community on the value of using inbred experimental animals, used to compare the special mice to pure chemicals. He suggested to his fellow scientists that just as they would not do chemical studies with undefined mixtures or impure substances, they should not try to do biological studies with unknown and varable mixtures of genes.

Dr. Seldon E. Bernstein, one of 35 staff scientists, notes that the mouse's value to research is further enhanced by its short reproductive cycle. It breeds at 50 to 60 days of age and produces a litter 20 days later, repeating the feat two or three times a year. "More generations of mice have lived since 1929 than of people since the human species evolved," Dr. Bernstein observed.

The mouse and the rat are highly unusual in their ability to inbreed. A new inbred strain of mouse takes about seven years to develop, a task that involves matings between brothers and sisters for 20 successive generations. At that point, it has been calculated, 99.999 percent of the genes in each animal are identical to those of its siblings.

An inbred strain of rabbit took the laboratory 35 years to produce. Other animals, including people, suffer from "inbreeding degeneration" when repeatedly mated with brothers and sisters. They become sickly and eventually fail to reproduce at all.

☐

Breeding is only a small part of the work at Jackson lab. Most of the staff scientists seek answers to human ailments that may be found among the mutant strains. Dr. Bernstein, for example, has figured out how to cure a genetically caused anemia in mice that mimics a rare human

stocks room. Although this mouse may not have a genetically caused deformity, the scientists here will try to find out by attempting repeated inbreeding of its offspring. If the defect is genetically caused—and the answer could take several years to determine—then biomedical scientists could have a new animal "model" that might aid research into the causes of human birth defects.

"We use inbred strains because they're a clean, pure test system," Priscilla Lane, a Jackson geneticist, explained. "We know that the effects we observe are due to the particular thing we're studying, whether it's a chemical carcinogen or environmental hazard or drug treatment." This greatly reduces the number of animals needed to recognize an effect. If all the animals in a study anemia called Blackfan-Diamond disease. He found that the mouse disease was caused by a defect in the bone marrow cells that produce red blood cells. The mouse can be cured by a transfusion or bone marrow transplant of these so-called stem cells. The studies suggest possible treatments for human anemias that may be caused by environmental agents such as drugs or radiation. Based on his studies, Dr. Bernstein has concluded that persons who work with radiation or toxic chemicals should have small amounts of bone marrow put in a "bank" as a medical insurance policy.

In the island resort town of Bar Harbor, where fog- and tourist-filled summer days yield to cold and lonely winters, the mouse has become the most important year-round resident. It is treated accordingly. Housed in rooms that are environmentally constant, the cages and bedding of pine and cedar shavings are changed weekly. They are fed made-to-order sterilized feed, and their air is changed 10 times an hour.

All who enter the animal rooms must remove street clothes and don carefully laundered "mouse suits." To further reduce the possibility of disastrous epidemics, the cages are covered by filters that prevent infectious organisms from traveling from cage to cage. Caretakers are trained to spot not only possible new mutants but any animal that may appear even a little bit sick. All suspected of harboring disease are sent with their cage mates to a health inspector for examination and quarantine.

The precious animals, stacked seven cages high, occupy a total of 72,000 square feet, and a typical mouse room is 1,500 square feet, comparable to the floor space in an average three-bedroom house. The nonprofit laboratory sells two-thirds of the mice it raises to researchers in every state and many foreign countries. Prices range from $2 each for the stripped-down model of a normal

inbred mouse that breeds easily to $20.60 each for the difficult-to-breed mouse with muscular dystrophy.

The dystrophic mouse, among other Jackson mutants with counterparts to severe human ailments, does not reproduce itself. To create one, geneticists mate two normal mice that are carriers of the gene for dystrophy. One quarter of their offspring will have the double dose of the dystrophy gene and show the symptoms of the muscle-wasting disease.

To improve the reproductive odds, the scientists can transplant the ovaries from a dystrophic mouse into a normal female of the same inbred strain. When mated with a carrier male, this female will produce half her offspring with dystrophy. In another approach, the female with the ovarian transplants can be artificially inseminated with semen from a dystrophic male, resulting in 100 percent dystrophic off-spring.

Even if no one is in the immediate market for a dystrophic mouse, the lab has to keep the mutant gene going for future use. This requires continual breeding of known carriers of the dystrophy gene, a time- and space-consuming enterprise. Therefore, lab scientists have begun to look into the feasibility of storing desired genes by freezing embryos that harbor them. When a particular gene is needed for research, the embryo is defrosted and transplanted into a foster mother to nurture until weaning.

Dr. George D. Snell, retired senior scientist of the laboratory, believes the value of the inbred mouse will continue to grow, especially with the mushrooming need to test chemicals and contaminants for their ability to cause toxic effects and genetic damage. The growing interest in transplantation of organs is a further stimulus, the mouse having already guided much of the current understanding of tissue incompatibility between unrelated persons.

Plant-Animal Interaction

by Jane E. Brody

The thermometer on the ground next to us read 152 degrees Fahrenheit as we lay on our bellies in the midafternoon desert sun, trying to unravel the social structure of a newly discovered species of seed-harvesting ant. But the ants were not so foolish as their two-legged observers. With one or two exceptions, they waited out the scorching heat in the relatively cool tunnels of their huge underground nest.

Then, about an hour after nightfall when the temperature of the sandy topsoil dropped to a more comfortable 90 degrees, the pace of activity picked up. We counted the workers that carried in the tiny seeds that the ants store and live on, and we counted the workers that carried out the chaff, which litters the desert floor. Unseen below ground, other workers separated the seed from the chaff.

Winged forms—believed to be males and females ready for mating—also appeared with increasing frequency as the night wore on. And occasionally there emerged a "major worker," twice the size of the others, to clear a nest opening of accumulated debris. At 10:30 P.M., necks aching from our flashlight vigil, we called it quits,

although the ants undoubtedly worked on through the cool of the star-ridden night.

This study, one of many to cross the frontiers of desert science, is part of an ambitious effort to unravel the complex relationships between the plants and animals that make the unforgiving environment of Baja's Central Desert their home. Through periodic sojourns during the last several years, researchers from Idaho have discovered in the Baja desert more than a dozen animal species new to science, including a bi-valve crustacean that lives in desert rainwater pools, various beetles and ants, and a new genus of a quarter-inch-long pseudoscorpion.

They have also discovered new associations between plants and animals. For example, the top of the barrel cactus was found to have nectar factories outside its flowers; stinging ants feed on the nectar and in return are believed to protect the cactus fruits from being eaten by bugs before the plant can seed itself.

Some 800 miles to the north, near Reno, a similar project proceeds in the Great Basin Desert of Nevada, the last major area of the contiguous United States to be explored. The harshness of this shrubby, treeless and parched terrain tested the mettle of gold- and silver-hungry miners and pioneer families who in the mid-1800's sought their fortunes in California. The wagon trails across the desert are littered with the bones of oxen that succumbed to starvation and dehydration in the dry heat, or that perished in the sudden flash floods, sandstorms or snow storms for which this area is infamous.

Both projects are being funded by a private organization called Earthwatch of Belmont, Mass., and aided by volunteers from 16 to 65 who pay for the privilege of participating in field research. Although the Nevada and Baja studies are both basic research projects, their findings are expected to help guide human incursions into the desert, enabling man to mine the desert's resources with minimal disturbances to its delicate balance of life.

□

Contrary to popular impressions of the desert as a barren, lifeless, inhospitable wasteland, these studies have shown that a field guide's worth of plants and animals share one of the world's most fragile ecosystems. While not "teeming" with the colorfully obvious life-forms that are found in a tropical rain forest, the desert is host to a surprisingly large number of species that observers can easily find if they stop to look. Through evolution and survival of the fittest, each has found some way to cope with the deprivations of the desert, the primary one being lack of water. The Baja desert receives an average of four inches of rainfall a year, supplemented by early morning dew. The Great Basin areas under study get perhaps five inches of precipitation, but little or no dew.

Not much is known about the ecology of the desert because the environment challenges the tolerance of human observers and because funding organizations have been as unimpressed with this "nonproductive" land as the average citizen. Quick in-and-out collecting trips are not enough. To study interrelationships between plants and animals, researchers must spend weeks or longer in one spot, importing all food and water from home or resupplying from scattered small towns.

Patricia and Hamilton Vreeland of the University of Nevada at Reno said that the desert they are studying is a hotbed of geothermal energy and

mineral resources waiting to be tapped, located in the nation's fastest growing state. "It's not a pristine environment untouched by humanity, and it doesn't house unique species that we know of," Mr. Vreeland noted. "We're trying to collect baseline data from which one can evaluate the effects of further human intrusions," he continued. "We're not idealistic ecologists who say the area must be preserved as is just because it's here. We recognize the need for trade-offs. But we think one should know the costs of development before it's too late."

Just the effect of human visitors and their dune buggies can be devastating to the desert. The thin layer of desert soil is protected from erosion by a fine crust composed of fungi, algae and mosses. When this crust is broken by a wheel or foot, the soil can easily erode, bringing an end to the desert's ability to support life. Mr. Vreeland continued, "The desert is fragile because, compared to a Maine forest, it houses a limited number of species." A disturbance in one can bring an end to another. Also, his co-worker Tom Lugaski added, "You can go out and do a lot of damage in a short time, and then it takes a very long time to undo that damage."

In the Baja, the desert workday begins at 5:30 A.M., when the sky lightens but before the scorching sun rises. Animal traps are checked, their inhabitants carefully measured and recorded and then released. Birds and bats caught in fine nets are similarly measured and stuffed for further study. Using hoola hoops to mark the study areas, an accounting is made of the plants, animal droppings and human litter found along the right-of-way cut five years ago for Baja's transpeninsular highway, a two-lane, shoulderless road with frequent washouts and more and bigger potholes than all of New York City's streets put together.

Despite its condition, the road has brought tourists to this otherwise inaccessible desert; on one evening, the research team counted several dozen recreational vehicles passing through, carrying people and objects that could disrupt this fragile environment.

After a break for breakfast at 8:30, work resumes until lunch, after which the heat—100-plus in the shade, 150 on the sunny ground—usually precludes physically taxing work until dusk. The hot afternoons are a time for napping in the shade of the desert's huge granite

Small nectar factories discovered on the barrel cactus attract at least eleven different ant species. The ants collect nectar from the cactus and in return may discourage plant-eating insects from feeding on the cactus.

Expedition leader William H. Clark and the author check the height of a cardon cactus. Large cardon specimens can hold a half-ton of water.

boulders, reading under the parachute that serves as a parasol over the campsite, pinning insects or stuffing animals. Later, the researchers take advantage of the evening cool, excavating ant nests, tracking down scorpions with a black light that causes them to luminesce bright green on the otherwise dark desert floor, hunting long-horned beetles that feed only at night.

"There are three basic ways to adapt to the desert," explained Dr. Robert Bratz of the College of Idaho, a longtime student of the Mexican desert who co-ordinated the Baja expedition directed by William H. Clark. "You learn how to get more water, you reduce the loss of water or you store water." Animals like coyotes and birds can travel to the desert's infrequent water holes.

But small mammals must make do with what water happens to be nearby. The wood rat, for example, chews its way into the watery chambers of the massive cardón cactus, which then forms a thick scar tissue over the wood rat tunnels to protect itself from further dehydration. The cardón resembles Arizona's saguaro and, along with a bizarre-looking woody plant called the boojum, dominates the Baja desert landscape.

Other rodents stay underground during the heat of day, emerging only at night to gather food. The fleshy roots and tubers of many desert plants keep some desert animals supplied with water. Jackrabbits and cottontails seek their meals at dawn and dusk. Lizards and snakes have scaly skins and insects have outer skeletons that keep water loss to a minimum. Even then, few are to be seen when the sun is high in the sky. Mr. Lugaski explained that the lizard partly controls its body temperature by changing color. It starts out dark-skinned in the morning; then as the temperature rises, pigmented structures in the skin close up and the skin lightens to reflect more heat. With nightfall, the skin again turns dark.

Perhaps the best-adapted desert mammal, Dr. Bratz noted, is the kangaroo rat (really a mouse), which can survive in adult life entirely on dry seeds. It excretes only solid wastes and can derive all the water it needs from the metabolic breakdown of fats and carbohydrates in the seeds. This long-tailed rodent, along with the smaller pocket mouse, cleaned the campsite of crumbs every night.

Plants have a somewhat harder time in the desert, since they are destined to stay where they happen to germinate. The cactus is nature's answer to water conservation. Its leaves are reduced to spines and it makes food through photosynthesis in its thick green stem. The innards of the cactus contain gelatin-like substances that attract and hold water—up to half a ton in the big cardóns, which are accordion-pleated for easy expansion when nature provides a surfeit of water. Most cactuses sport a massive network of surface roots, which act as catchbasins to retain every drop of available rainfall.

The boojum, named for the mystical thing in Lewis Carroll's "The Hunting of the Snark," is in essence a water tower—a woody, tapering cylinder filled with a bitter liquid. This tree, which is not a cactus and grows naturally only in the Baja Peninsula, resembles an upside-down carrot, with short scraggly branches extending from the trunk. Its close relative, the ocotillo, also known as coach whip or monkey tail cactus, leafs out only after a rain and photosynthesizes for a while; then the leaves drop off when the environment dries up again.

Other plants, like the mesquite, a bush-like tree under which Mexican cattle often seek relief from the sun, put down very deep taproots to reach the desert water table, perhaps 200 feet below the sandy surface. The seeds of many desert plants are coated with chemical inhibitors; they can germinate only when rain washes the chemical off. Then it is a rush through the life cycle—from sprout to bloom to fruit—before the water supply is exhausted.

Patricia Vreeland said that during the first Earthwatch expedition into the Great Basin Desert in June 1976, four inches of snow (representing half that area's average annual precipitation) collapsed several tents and turned the parched hillsides into rivers and the desiccated valleys into impassable mudholes. But a week later the desert was a blaze of color as desert annuals cashed in on the unexpected windfall of water.

As Dr. Bratz pointed out, "There are no seasons, as such, in the desert. The desert blooms whenever it happens to rain." This fact makes desert study a rather chancy enterprise, since one never knows for sure which plants and animals will be available for study. "We always start out with far more projects in mind than we could possibly do in the allotted time because all aren't going to be available," Mr. Clark said. "Just because a certain plant is *supposed* to flower in June doesn't mean it will, and then the insect that feeds on it won't be there for study." He added that a field biologist "also has to be prepared for the unexpected—for example, the opportunity to study a plant or animal that just happens to be around when you are." It is this kind of unpredictability that distinguishes field from laboratory research. In the lab, Mr. Clark said, everything can be carefully controlled—"you come in in the morning, find a blue bottle on the table, mix it with the pink bottle and get orange."

In the desert that the Vreelands and Mr. Lugaski study, there are no notable cactuses or succulents as in the Baja. Here, foot-high silvery-green shrubs mark the lowlands—mostly sagebrush and shadscale along with spiny hop-sage, rabbit brush, three-winged salt brush and tumbleweed (an imported species). Short evergreens—piñon pine and juniper—dot the cooler mountain slopes.

Yet the Great Basin Desert abounds with animal life—lizards, rabbits, squirrels, birds, desert mice and rats, and snakes, including the fierce panamint rattler as well as the more timid Great Basin rattler. A pair of great horned owls is nesting this year at one former Earthwatch site, and judging from the bones beneath their rock cliff, they dine well on small rodents and rabbits.

PHOTO ACKNOWLEDGMENTS

P. 26, 27, 29, 31, 32 Wide World Photos; p. 34-38, 40, 42-48 (top) Kenya Tourist Office; p. 48 (bottom) Lebrun E. Burnett/Kenya Tourist Office; p. 50 Australian Information Service; p. 52, 53 Arizona Office of Tourism; p. 54 South African Consulate General; p. 56 The New York Times; p. 58, 60 Travel Marketing Section, State of Colorado; p. 62, 63 The New York Times; p. 68-74 U.S. Fish and Wildlife Service; p. 77 Walter Sullivan/The New York Times; p. 78 Robert Trumbell/The New York Times; p. 89 United Press International; p. 91 David Strick/The New York Times; p. 92, 95, 96 United Press International; p. 98 Susan Cottingham/Town of Crested Butte; p. 100 Wes Light/Town of Crested Butte; p. 102 The New York Times; p. 103 Henry Truebe/Town of Crested Butte; p. 104, 105, 106 (top) Wes Light/Town of Crested Butte;p. 106 (bottom) Crested Butte Archives; p 108 Debbie Hooks-Drake/Town of Crested Butte; p 111 J. Goerg/New York State Department of Environmental Conservation; p. 112 United Press International; p. 114 U.S. Environmental Protection Agency; p. 115 United Press International; p. 117 EPA-Documerica/Charles O'Rear; p. 120, 121 Jane E. Brody/The New York Times; p. 125 The Aluminum Association, Inc.; p. 126, 128, 130 Grumman Energy Systems, Inc.; p. 131 Schwinn Bicycle Company; p. 135 Food and Drug Administration; p. 136 Dr. Irving J. Selikoff; p. 137 The American Cancer Society; p. 139 United Press International; p. 142 Dr. David Etnier/U.S. Fish and Wildlife Service; p. 143 Don Rimbach/U.S. Fish and Wildlife Service; p. 144 Jerry A. Powell/U.S. Fish and Wildlife Service; p. 145 U.S. Fish and Wildlife Service; p. 146 Michael L. Smith/U.S. Fish and WIldlife Service; p. 147 Glen Smart/U.S. Fish and Wildlife Service; p. 148-149 Robert Conrad/The New York Times; p. 150 Luther C. Goldman/U.S. Fish and Wildlife Service; p. 152, 154, 155, 156, 158, 160, 162 DuPont; p. 164-171 Exxon Corporation and Jack Van Hyning, Nancy Simmerman, Steve McCutcheon, Alaska Pictorial Service, George Watson, Leo Touchet, Le Photo Circle; p. 174, 175 The New York Times; p. 176-181 courtesy of Deepsea Ventures Inc.; p. 183 Teresa Zabala/The New York Times; p. 188 The New York Times; p. 191 The New York Times; p. 196 Culver Pictures; p. 197 California Department of Fish and Game; p. 202, 203, 205 Boyce Rensberger/The New York Times; p. 206 The New York Times; p. 210 Anthony Wolff/The New York Times; p. 212 Rodica Prato/The New York Times; p. 217 The New York Times; p. 218, 220 National Oceanic and Atmospheric Administration; p. 221 David O'Neill/The New York Times; p. 223-225 National Oceanic and Atmospheric Administration; p. 227-229 Lawrence K. Altman/The New York Times; p. 230, 231 The Mayo Clinic; p. 234 Malcom W. Browne/The New York Times; p. 236 Dr. William N. Lipscomb/Harvard University; p. 239, 240 Phil Greer/The New York Times; p. 243 (top), 244 Bill Dupuy/Jackson Laboratory; p. 243 (bottom) George McKay/Jackson Laboratory; p. 248 (left) Mary H. Clark; p. 248 (right) William H. Clark

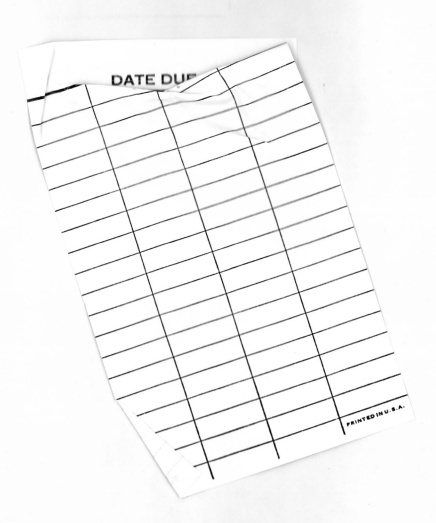

DATE DUE

PRINTED IN U.S.A.